T0220419

Provenance:
An Introduction to PROV

Synthesis Lectures on the Semantic Web: Theory and Technology

Editors
James Hendler, *Rensselaer Polytechnic Institute*
Ying Ding, *Indiana University*

Synthesis Lectures on the Semantic Web: Theory and Application is edited by James Hendler of Rensselaer Polytechnic Institute. Whether you call it the Semantic Web, Linked Data, or Web 3.0, a new generation of Web technologies is offering major advances in the evolution of the World Wide Web. As the first generation of this technology transitions out of the laboratory, new research is exploring how the growing Web of Data will change our world. While topics such as ontology-building and logics remain vital, new areas such as the use of semantics in Web search, the linking and use of open data on the Web, and future applications that will be supported by these technologies are becoming important research areas in their own right. Whether they be scientists, engineers or practitioners, Web users increasingly need to understand not just the new technologies of the Semantic Web, but to understand the principles by which those technologies work, and the best practices for assembling systems that integrate the different languages, resources, and functionalities that will be important in keeping the Web the rapidly expanding, and constantly changing, information space that has changed our lives.
Topics to be included:

- Semantic Web Principles from linked-data to ontology design

- Key Semantic Web technologies and algorithms

- Semantic Search and language technologies

- The Emerging "Web of Data" and its use in industry, government and university applications

- Trust, Social networking and collaboration technologies for the Semantic Web

- The economics of Semantic Web application adoption and use

- Publishing and Science on the Semantic Web

- Semantic Web in health care and life sciences

Provenance: An Introduction to PROV

Luc Moreau and Paul Groth

ISBN: 978-3-031-79449-0   paperback
ISBN: 978-3-031-79450-6   ebook

DOI 10.1007/978-3-031-79450-6

A Publication in the Springer series
SYNTHESIS LECTURES ON THE SEMANTIC WEB: THEORY AND TECHNOLOGY

Lecture #7
Series Editors: James Hendler, *Rensselaer Polytechnic Institute*
               Ying Ding, *Indiana University*
Series ISSN
Synthesis Lectures on the Semantic Web: Theory and Technology
Print 2160-4711   Electronic 2160-472X

Provenance:
An Introduction to PROV

Luc Moreau
University of Southampton

Paul Groth
VU University of Amsterdam

SYNTHESIS LECTURES ON THE SEMANTIC WEB: THEORY AND TECHNOLOGY #7

ABSTRACT

The World Wide Web is now deeply intertwined with our lives, and has become a catalyst for a data deluge, making vast amounts of data available online, at a click of a button. With Web 2.0, users are no longer passive consumers, but active publishers and curators of data. Hence, from science to food manufacturing, from data journalism to personal well-being, from social media to art, there is a strong interest in *provenance*, a description of what influenced an artifact, a data set, a document, a blog, or any resource on the Web and beyond. Provenance is a crucial piece of information that can help a consumer make a judgment as to whether something can be trusted. Provenance is no longer seen as a curiosity in art circles, but it is regarded as pragmatically, ethically, and methodologically crucial for our day-to-day data manipulation and curation activities on the Web.

Following the recent publication of the PROV standard for provenance on the Web, which the two authors actively help shape in the Provenance Working Group at the World Wide Web Consortium, this Synthesis lecture is a hands-on introduction to PROV aimed at Web and linked data professionals. By means of recipes, illustrations, a website at www.provbook.org, and tools, it guides practitioners through a variety of issues related to provenance: how to generate provenance, publish it on the Web, make it discoverable, and how to utilize it. Equipped with this knowledge, practictioners will be in a position to develop novel applications that can bring openness, trust, and accountability.

KEYWORDS

provenance, prov, audit trail, compliance, audit, traceability, semantic web

To Ryan, Sammey, and Achika — L.M.

To Marieke — P.G.

Contents

Preface

This book stems from the authors' decade of experience related to provenance, their leadership of the W3C Provenance Working Group, and several tutorials on provenance delivered at FIS'10,[1] IPAW'12,[2] ISWC'12,[3] and ESWC'13.[4] The PROV specifications provide normative and non-normative material about data models and protocols related to provenance, but offer very little guidance on how to design and deploy provenance. The purpose of this book is to address this concern, by providing a compact and practical guide to developing PROV-based applications, combined with a fully-deployed example that readers can inspect and study in detail.

This book is intended for developers aiming to make their systems provenance-aware. It can also be used as a textbook for an undergraduate course on provenance, or potentially part of a larger module on the Semantic Web or Information Assurance.

We assume that readers are familiar with core technologies of the Web and Semantic Web. Specifically, understanding of URIs, basic understanding of HTTP, basic notions of HTML, and the principles of RDF are prerequisites for using this book.

The rest of this book is structured as follows.

- Chapter 1 provides an introduction to provenance.

- Chapter 2, *Data Journalism Scenario*, introduces a data journalism example, which we use extensively across the book. It describes the publication of an article by a fictitious news agency, based on some government statistics recently published. This example is used to illustrate many use cases that can be addressed by means of provenance. The chapter also provides a brief introduction to how provenance can be represented.

- Chapter 3, *The PROV Ontology*, is a concise presentation of the PROV ontology, which can be used as reference material for this book.

- Chapter 4, *Provenance Recipes*, is concerned with methodological recipes on how to model provenance for specific problems, and how to deploy it in an inter-operable manner.

- Chapter 5, *Validation, Compliance, Quality, Replay*, first focuses on the notion of valid provenance. It then expands on various forms of utilization of provenance. A series of technical requirements is introduced, and SPARQL queries are used to illustrate how these can be implemented.

---

[1]OPM Tutorial: http://openprovenance.org/tutorial/
[2]PROV Tutorial at IPAW'12: http://www.w3.org/2011/prov/wiki/IPAW_2012_Tutorial
[3]PROV Tutorial at ISWC'12: http://www.w3.org/2011/prov/wiki/ISWCProvTutorial
[4]PROV Tutorial at ESWC'13: http://www.w3.org/2001/sw/wiki/ESWC2013ProvTutorial

- Chapter 6, *Provenance Management*, is dedicated to techniques to manage provenance, and specifically to make it available, by means of RDFa embedded in HTML documents and provenance services. A series of libraries, services, and tools are then briefly discussed. Finally, a guided tour of http://www.provbook.org illustrates the provenance management techniques deployed on the website associated with the book.

- Chapter 7, *Conclusion*, summarizes the book with a checklist that developers can follow to check whether their provenance is properly structured and exposed. Open issues and future research directions are also discussed.

Luc Moreau and Paul Groth
August 2013

Acknowledgments

Luc Moreau wishes to thank Alex Fraser, Trung Dong Huynh, Mike Jewell, Amir Sezavar Keshavarz, and Danius Michaelides for their work on the Southampton Provenance Tool Suite.

Paul Groth thanks Frank van Harmelen for supporting his involvement in the working group and Stefan Schlobach for putting up with constant telecons.

The authors thank the entire W3C Provenance Working Group for their efforts producing PROV, and Trung Dong Huynh for his comments on a draft of the book. Additionally, the authors thank Yolanda Gil for her work leading to the creation of the W3C Provenance Working Group.

Luc Moreau's work is supported under SOCIAM: The Theory and Practice of Social Machines; ORCHID: Human-Agent Collectives: From Foundations to Applications; Smart-Society: hybrid and diversity-aware collective adaptive systems: where people meet machines to build a smarter society; and PATINA: Personal Architectonics Through INteractions with Artefacts.

- The SOCIAM Project is funded by the UK Engineering and Physical Sciences Research Council (EPSRC) under grant number EP/J017728/1.

- The ORCHID Project is funded by the UK Engineering and Physical Sciences Research Council (EPSRC) under grant number EP/I011587/1.

- The SmartSociety Project is funded under FP7 Grant agreement number 600854.

- The PATINA Project is funded by the UK Engineering and Physical Sciences Research Council (EPSRC) under grant number EP/H042806/1.

Paul Groth's work is supported by the Data2Semantics project in the Dutch national program COMMIT as well as the EU IMI Open PHACTS project. Open PHACTS receives financial support provided by the IMI-JU, grant agreement number 115191.

Luc Moreau and Paul Groth
August 2013

CHAPTER 1

Introduction

In the world of art, the notion of provenance is well understood. A piece of art sold in an auction is typically accompanied by a paper trail, documenting the chain of ownership of this artifact, from its creation by the artist to the auction. This documentation is referred to as the provenance of the artifact. Provenance allows experts to ascertain the authenticity of the artifact, which in turn influences its price. If provenance is lost or incomplete, the artifact is likely to be auctioned for a lower price. In their book titled *Provenance*, Salisbury and Sujo [35] paraphrase the appreciation of provenance by a purveyor of art: "The more prestigious or infamous the previous owner, the better. A piece of art with a juicy history was always worth an extra ten grand." They further quote Werner Muensterberger (*Collecting: An Unruly Passion*), who provides a collector's perspective on provenance: "Buying a painting that was once owned by a well-known person means, in a way, standing in their shoes, walking in their footsteps, possessing a small part of their myth."

Whether leading to financial gain, establishing authenticity, or giving the illusion of belonging to a famous history, provenance is highly regarded in artistic circles, and as this chapter shows, is appreciated well beyond them. But on the Web, or generally, in large-scale, multi-stakeholder, or fast-processing systems, papers trails are not tractable: a form of provenance that can be representable in computer systems and processed by them is crucial. This is the purpose of this book: to introduce a new *standard model of provenance*, PROV, that can be applied to anything, not just pieces of art, but also to food and to supply chains, and importantly to data in computer systems. Its applications are multi-faceted and include data journalism, the Social Web, science, and business applications.

1.1 THE CASE FOR PROVENANCE

Food Provenance   The recent horsemeat scandal in Europe[1] is a reminder that the supply chain is critical in determining the quality of all food products. When horsemeat was discovered in beef products, the ability to trace back through the chain of suppliers, food processing plants, slaughter houses, and original farmers was essential in determining at which point mislabelling took place. Thus, provenance is particularly useful in problematic situations to audit complex processes that involve many stakeholders. Furthermore, a computer-based representation and processing of provenance is crucial to deal with the vast quantities of information, and the timeliness requested by both the public and regulatory authorities.

[1]http://www.bbc.co.uk/news/uk-21393180

Vice versa, good provenance is being turned into a competitive advantage by many food suppliers: fair trade products, appelations d'Origine Contrôlée, carbon neutral goods, sustainable productions[2] are examples, in which the origin of products and the quality of their manufacturing processes are an explicit part of their branding, to distinguish them from the competition.

Open Data and Data Journalism   Data journalism is a journalistic process whose purpose is to produce news based on open data or, more generally, any digital information. An ethos of data journalism is to expose the data and methods used to produce news items [13]; for this reason, data journalism is not only a consumer, but also a producer of open data. To achieve this aim, data can be massaged, mashed up, and republished. But, as noted by Chambers and Keegan,[3] since data wrangling can introduce errors, data journalists should care about the validity of data; thus, they recommend that provenance of data should include its primary source, but also all the transformational steps performed by anyone.

Since these steps are expected to be reproducible, such a kind of provenance empowers readers to validate results, analyze data with other parameters, publish their results, and become data journalists themselves, overall contributing to a more transparent and accountable environment.

Tracing Information in the Social Web   The first generation of the Web put users in a role of information consumers, whereas its successor, the Social Web, greatly facilitates a role of information curator.[4] In the Social Web, users are given the ability to select contents from across the Web, integrate it, edit it, rate it, publish it, and share it with others. This consume-select-curate-share workflow is similar to the data wrangling performed by data journalists, but typically requires very little technical skills, since it is supported by sites or applications such as Facebook or Flipboard.

Furthermore, on the Social Web, people are not the only curators: computer programs also take this crucial role. The collectives formed by the dynamic assembling of users and computers is often referred to as *social machines* [4].[567] Spectacular examples of such machines include crowd-sourcing applications such as Wikipedia[8] and Ushahidi,[9] which involved thousands of participating users, devices, and bots, to produce an encyclopedia and detailed maps of Haiti, respectively, in relatively short timespans. As enrolled users and machines contribute contents, the inevitable question of trust comes into play. Why should we trust information on the Social Web? Jarvis[10] comments on the importance of provenance in the context of the Web, where there is a shift to

---

[2]Ethical eating: http://www.greenhotelier.org/our-themes/community-communication-engagement/the-rise-of-ethical-eating-provenance-on-a-plate/

[3]Data Journalism: http://datadrivenjournalism.net/featured_projects/how_spending_stories_spots_errors_in_public_spending

[4]Flipboard: http://www.guardian.co.uk/technology/appsblog/2013/mar/27/flipboard-2-curating-digital-magazines

[5]Orchid: http://www.orchid.ac.uk

[6]Sociam: http://www.sociam.org

[7]SmartSociety: http://www.smart-society-project.eu

[8]Wikipedia: http://www.wikipedia.com

[9]Ushahidi: http://www.ushahidi.com

[10]The importance of provenance: http://buzzmachine.com/2010/06/27/the-importance-of-provenance/

curation: "Good curation demands good provenance." He continues: "Provenance is no longer merely the nicety of artists, academics, and wine makers. It is an ethic we expect." Provenance, which indicates who contributed to information helps consumers check where information comes from, why it was selected, and how it was edited.

Reproducibility of Science   As the e-science vision becomes reality [20, 21], researchers in the scientific community are increasingly perceived as providers of online data, which take the form of raw data sets from sensors and instruments, data products produced by workflow-based intensive computations, or databases resulting from curation. While science is becoming computation and data intensive, the fundamental tenet of the scientific method remains unchanged: experimental results need to be reproducible. In contrast to a workflow [12], which can be viewed as a recipe that can be applied in the future, provenance is regarded as the equivalent of a logbook, capturing all the steps that were involved in the actual derivation of a result, and which could be used to replay the execution that led to that result, so as to validate it.

Accountability, Transparency, Compliance in Business Applications   Steve New[11] refers to the provenance of a company's products, and explains how businesses have changed their practices to make their supply chains transparent, because they worry about quality, safety, ethics, and environmental impact. Fox[12] observes that governments increasingly request transparency and provenance information in the area of anti-corruption compliance. Generally, there are many regulatory frameworks that request institutions to be able to trace back their processes, and demonstrate that events happened the way they should have. As Weitzner [38] notes, provenance is a substrate that can be used to perform policy checks and to make systems accountable.

1.2   A DEFINITION OF PROVENANCE

In this book, we adopt the W3C (World Wide Web Consortium) Provenance Working Group's definition of provenance:[13]

> Provenance is defined as a record that describes the people, institutions, entities, and activities involved in producing, influencing, or delivering a piece of data or a thing.

In the context of the Web, provenance is a record that can be created by, exchanged between, and processed by computers. This record contains descriptions of events that led to a piece of data or a thing being in a given state: thus, provenance can pertain to documents, data, or more generally resources over the Web, but also to things in real or even imaginary worlds.

The computer-processable record contains descriptions of the events that took place, leading to a resource or a thing, as it exists in some context. Many factors can contribute to such a state

---

[11]Transparent supply chain: `http://hbr.org/2010/10/the-transparent-supply-chain/ar/1`
[12]Provenance in the supply chain: `http://www.lexisnexis.com/community/corpsec/blogs/corporateandsecuritieslawblog/archive/2010/10/21/provenance-in-the-supply-chain-transparency-and-accountability-under-the-fcpa-and-bribery-act.aspx`
[13]Definition of provenance: `http://www.w3.org/TR/2013/REC-prov-dm-20130430/#dfn-provenance`

of affairs, including the people involved, the organizations they act on behalf of, the processes that are being executed, and other data, resources, or things that play a part in it. Conceptually, this computer-processable record takes the shape of a graph consisting of nodes (for the people, organizations, data, things, and processes) and edges (expressing how they relate to each other). For this reason, we will use the term *provenance graph* when we wish to refer the graphical nature of the provenance record, but will also use *provenance trail* and *provenance description* as synonyms.

1.3   PROVENANCE AND THE WEB ARCHITECTURE

Tim Berners-Lee, the inventor of the World Wide Web, envisioned[14] a browser button by which the user can express uncertainty about a document being displayed: "So how do I know I can trust this information?". Upon activation of the button, the software then retrieves metadata about the document, listing assumptions on which trust can be based. To be able to achieve such a vision, provenance needs to have a representation, and needs to be accessible in the context of the Web.

Provenance records can be expressed and encoded using Web technologies, so that they can be exchanged across the Web, and exploited by users and services. A natural question is where provenance fits in the Web architecture, and particularly, in the Semantic Web architecture. While we recognize its limitation, the Semantic Web layer cake diagram [3] remains an architectural reference. In Figure 1.1, we present a variant of the diagram that incorporates a provenance layer.

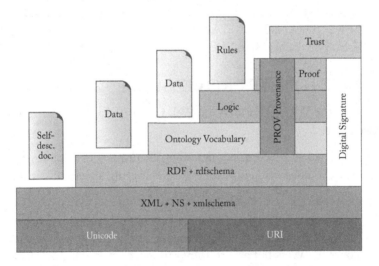

Figure 1.1: Provenance in the Semantic Web Layer Cake Diagram.

The layers of the Semantic Web layers can be summarized as follows (Figure 1.1). Unicode is the standard that allows people use computers in any language, whereas Uniform Resource

[14]Oh Yeah? button: http://www.w3.org/DesignIssues/UI.html#OhYeah

Identifiers (URIs) are the mechanism to identify resources in the Web Architecture. XML is the markup language to encode documents and data in both machine and human readable ways. RDF (Resource Description Framework) allows for the description of resources, whereas ontologies are capable of specifying things and relationships between them. The logic layer allows for derivation of new knowledge from assertions published on the Web. Proofs are the result of keeping track of logical inferences. Cryptographic techniques and in particular digital signatures, can be used for authentication and non-repudiation across the Web. And, finally, trust may be established using such proofs and cryptographic techniques.

PROV is the new standard for provenance defined by the World Wide Web Consortium. PROV spans multiple layers: ontology vocabulary, logic, and proof. PROV consists of an ontology PROV-O [27] expressed in the OWL2 Web Ontology Language. PROV assertions, compliant with the PROV ontology, can be expressed and published on the Web. PROV defines inference rules [8] allowing domain independent reasoning about provenance assertions. Having expressed and reasoned over provenance assertions about the origin of data, everything is in place to make judgments about the reliability or trustworthiness of data: hence, provenance is a key component to support trust on the Web.

Figure 1.1 provides a blueprint for a set of protocols, data formats, and knowledge representation techniques for the Semantic Web developer. This diagram should be interpreted with some flexibility. Indeed, not all logical reasoning requires ontologies, and other data formats, such as JSON, are also frequently encountered over the Web. But the essence of provenance, as a vehicle to establish trust on the Web, remains, whatever variant of the layered diagram is considered.

1.4   THE W3C PROV STANDARD

The W3C Provenance Working Group has published 13 documents related to provenance. (For a complete reference to all the specifications, see PROV-OVERVIEW [19].) This book makes extensive use of six specifications, which are organized as per Figure 1.2.

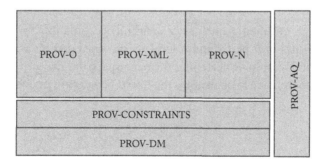

Figure 1.2: Simplified view of PROV specifications (inspired by [19]). Specifications marked in orange are normative (W3C Recommendations), whereas those in blue are informative (W3C Notes).

Since provenance is intended to describe the flow of data across multiple systems, it is crucial for provenance to be independent of the technologies used in those systems' executions. Furthermore, those systems are likely to be heterogeneous, implemented by different developers or companies, and may each have their own way of representing information. Hence, the PROV standard for provenance adopts the idea of a *conceptual data model* (PROV-DM [33]), which can be serialized in multiple formats. Three ways of serializing PROV are defined by the W3C. The PROV ontology is an OWL2 ontology allowing the mapping of the PROV data model to RDF [27], which itself offers multiple serialization formats, such as TURTLE and RDF/XML. PROV-XML is an XML schema for the PROV data model [22], allowing native XML representations. PROV-N is a textual notation for provenance designed for human consumption [34].

The PROV data model and its serializations essentially define a vocabulary, allowing provenance descriptions to be expressed in multiple formats. Provenance descriptions are said to be valid if they describe a consistent history of events: the notion of validity is defined in PROV-CONSTRAINTS [8].

Finally, once provenance has been expressed according to the PROV vocabulary, some conventions are required for publishers to make it available, and for consumers to discover and retrieve it. These conventions are specified in PROV-AQ [25].

1.5   ONLINE EXTENSIONS

One of the key drivers for this book is to make a realistic example of provenance available online. The data journalism example and all materials related to this book can be a found at http://www.provbook.org. Where possible this book uses URLs to the online site. Readers are strongly invited to follow these links, and study the contents, since they are a live example of the good provenance practice this book aims to advocate.

In this book, we rely on various namespaces identified by the following URIs and denoted by corresponding prefixes. The first part of the table lists standardized namespaces and namespaces for generic concepts of this book. The second part enumerates application-specific namespaces used in the book.

We further adopt qualified names as a convenience notation for URIs. For instance, nowpeople:Alice consists of a prefix and a local name, separated by a colon. From a qualified name, one can derive a URI by concatening the namespace URI denoted by the prefix and the local name. Hence, nowpeople:Alice denotes the URI http://www.provbook.org/nownews/people/Alice.

prefix	namespace URI	description
prov	`http://www.w3.org/ns/prov#`	The PROV namespace
xsd	`http://www.w3.org/2001/XMLSchema#`	The XML schema namespace
dct	`http://purl.org/dc/terms/#`	The Dublin Core Metadata Terms namespace
rdfs	`http://www.w3.org/2000/01/rdf-schema #`	RDF Schema
bk	`http://www.provbook.org/ns/#`	The namespace for terms defined in this book
now	`http://www.provbook.org/nownews/`	The namespace for NowNews
nowpeople	`http://www.provbook.org/nownews/people/`	The namespace for NowNews people
is	`http://www.provbook.org/nownews/is/`	The namespace for NowNews instance space
other	`http://www.provbook.org/othernews/`	The namespace for OtherNews
policy	`http://www.provbook.org/policyorg/`	The namespace for PolicyOrg
gov	`http://www.provbook.org/gov/`	The namespace for GovStat
govftp	`ftp://ftp.bls.gov/pub/special.requests/oes/`	The namespace for GovStat data
provapi	`http://www.provbook.org/provapi/documents/`	The namespace for provenance resources

Figure 1.3: Namespace prefixes and URIS.

CHAPTER 2

A Data Journalism Scenario

Chapter 1 gave a brief overview of how provenance can be applied to applications ranging from compliance in business to reproducibility in science. In this chapter, we delve into one scenario, *data journalism*, in more depth and provide use cases where provenance can be useful. Throughout the rest of the book, we will be revisiting this scenario to show concretely how the use cases presented here can be tackled through the application of provenance technologies.

Data Journalism [13] is the production of news based on digital information. It is often associated with infographics and visualisations of data products that are used to provide a summary of complex data to support the journalist's story. We selected this domain as an illustrative example for a number of reasons. Data journalism reflects the kinds of complex distributed production pipelines that are increasingly common in multiple fields. Information is furnished by both individuals in the organization as well as freelance or other independent providers and collated along multiple steps before a final outcome is seen. It is rare to see a report or output from an organization that has not been touched by multiple hands. Furthermore, data journalism highlights the sort of combination of automated and manual processes that are coming to characterise today's organizational outputs. Data journalism was also identified by the W3C Provenance Incubator Group[1] as a scenario that encompasses a wide variety of provenance use cases [15]. Finally, journalism is something that we all have some experience with and should make the scenario easier to understand.

The rest of this chapter is organized as follows. We first present a data journalism scenario and introduce the characters and organizations involved and their task of producing an article and associated report. We then discuss a number of use cases from this scenario that present challenges faced where provenance could be applicable. Finally, we introduce the core concepts of provenance illustrating them with respect to data journalism.

2.1 SCENARIO: THE EMPLOYMENT REPORT

NowNews is an online blog that provides up-to-the-minute news on a wide variety of public policy and business stories with the occasional entertainment story to lighten things up. NowNews prides itself on its data-driven approach as one of its unique selling points and has a number of in-house approaches to this end. Given both its data-driven and up-to-the-minute reporting,

---

[1]The W3C Provenance Incubator Group was established to look at the state of the art and develop a roadmap toward potential standardization. It was the precursor to the W3C Provenance Working Group. See http://www.w3.org/2005/Incubator/prov/wiki/W3C_Provenance_Incubator_Group_Wiki.

NowNews stories are often reused as a source of information by public policy organizations. In this scenario, we walk through the production and publication of a story by NowNews and its subsequent use by a public policy organization for a report. Let us first introduce the characters in the story.

2.1.1   CHARACTERS

Within the scenario there are two types of characters: organizations and people. Each character is given an identifier in the form of a URL. There are four key organizations:

1. NowNews - an online news blog focused on public policy and business (`now:NowNews`);

2. OtherNews - a traditional national newspaper (`other:OtherNews`);

3. PolicyOrg - a public policy organization creating reports designed for law makers and governmental organizations (`pol:PolicyOrg`); and

4. GovStat - the official government statistics office (`gov:GovStat`).

Within NowNews, we have four main people responsible for creating the story and publishing it.

1. Bob - the *journalist* (`nowpeople:Bob`);

2. Alice - because NowNews is so data centric, Alice is employed as a full-time *data cruncher* who is responsible for acquiring and processing data sets (`nowpeople:Alice`);

3. Tom - the *editor* who checks stories and is responsible for approving them before publication (`nowpeople:Tom`); and

4. Nick - the *webmaster* who publishes stories on the blog and manages the website (`nowpeople:Nick`).

We break down the scenario[2] into three parts: story creation and publication, crunching data, and reusing the story. Figure 2.2 depicts the scenario in terms of relationships between organizations, while Figure 2.1 focuses on the flow of information between the people in NowNews.

2.1.2   STORY CREATION AND PUBLICATION

The scenario focuses on a release of data by GovStat on employment statistics per occupation on a state-by-state basis. This data gets regularly updated and NowNews wants to be the first to provide an easily digestible view of this large data set. The release of data by GovStat is modeled on the Occupational Employment Statistics data release done by US Bureau of Labor Statistics.[3]

---

[2]A full implementation of the scenario can be found online at http://www.provbook.org.
[3]US Occupational Employment Statistics: http://stats.bls.gov/oes/

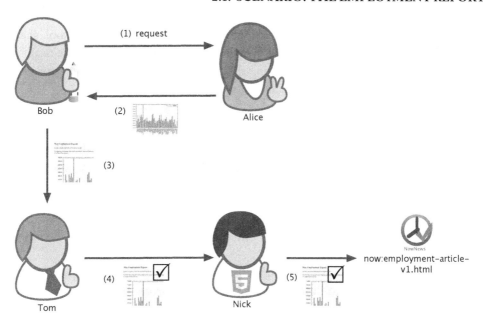

Figure 2.1: Illustration of how information flows between members of NowNews during story creation and publication in the Data Journalism scenario.

In a staff meeting, Bob is assigned to write an article about the release of the employment statistics that come out on March 27. Figure 2.1 shows the steps involved in preparing the article. (In this paragraph, we refer to the numbers in the figure.) To prepare the article, Bob asks Alice to download and create a chart of the data showing the mean salary across occupations per state across the country (1). While Alice produces the chart (discussed in the next section), Bob does some background work for the subsequent story and prepares some text based on it. In particular, Bob selects a quote to include in his article from an important economics blogger. After Alice supplies the chart (2), Bob embeds the chart in the story, and finishes writing the article. The story then goes through an editor's check (3) done by Tom before it is handed off to Nick (4), who prepares the article for publication on the Web (5). After publication, a subsequent revision of the economic figures was performed by GovStat. NowNews updates the article to reflect these revisions.

2.1.3 CRUNCHING DATA

Data journalism often requires significant data munging and processing to produce graphics that are compelling for readers. This case is no different, Alice first downloads the data from the GovStat website. It comes in a compressed zip file. After uncompressing the files, Alice has a large Excel file which she then converts to a csv file. To make extracting information easier and more reproducible, she then converts the data to RDF. Using a command-line tool, she executes a SPARQL query over the RDF to extract the mean salary per state. Alice plots the results in order to create the chart. When she is done creating the chart, she sends it over to Bob for inclusion in the article.

2.1.4   REUSING THE STORY

Because of its policy oriented approach, NowNews has many followers in non-governmental organizations, policy organizations, and other advisory firms. PolicyOrg is producing a report on employment across multiple states and in particular looking at the change in employment groupings over time (i.e., the move from manufacturing to services). In preparation of a report, they gather articles and evidence from multiple sources including the article produced by Bob as well as articles coming from OtherNews. They assemble these sources into a comprehensive report provided to policy makers.

In this scenario, we see how information is processed, aggregated and put together across multiple organizations to serve different consumer requirements. Using this scenario, we now illustrate a number of use cases where the provenance techniques discussed in the rest of this book can be applied to help address problems and make systems more effective.

2.2   PROVENANCE USE CASES

While the application of provenance technologies is wide, here, we focus on four use-case areas where provenance is particularly applicable: quality assessment, compliance, cataloging, and replay. In this section, our aim is not to discuss how provenance addresses these use cases but instead to lay them out. In the subsequent chapters, we will return to these use cases to see how provenance can be applied.

2.2.1   QUALITY ASSESSMENT

Users of content, in particular content that has been created through aggregation, need to be able to judge the quality of the content. Is the content trustworthy? Was it made with care? Are the people who produced it experts in the field? Was it produced from data or information that was inappropriately modified? The following are some specific cases in the context of data journalism where quality assessment is critical.

Use Case 2.1   The latest data - Timeliness   PolicyOrg is about to issue their report publicly. Before releasing the report they want to confirm that the report is based on the most up-to-date data. In particular, they have multiple charts, graphs and visualizations that are all based on multiple data sets. One of the figures that they have reused in the report stems from Bob's article on employment that appeared in NowNews. PolicyOrg needs to run a check that ascertains what data that figure was based upon.

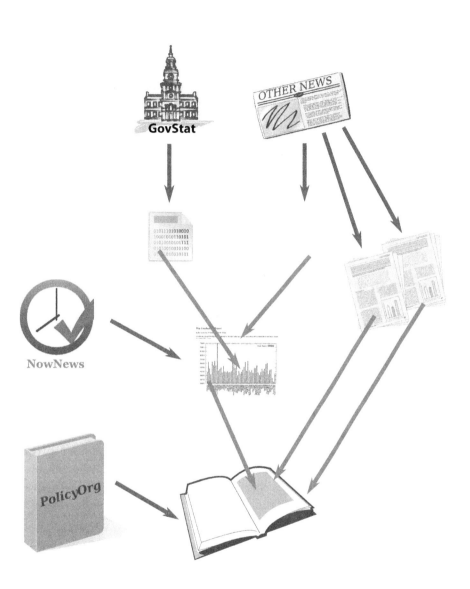

Figure 2.2: Illustration of the Data Journalism Scenario. Blue arrows indicate an organization responsible for the publication of an article, data set, or report; red arrows show how some artifact is used (by insertion or by processing) to produce another.

Use Case 2.2   Finding trusted articles   When putting together a story or a report, content creators want to find information that is based on trusted sources. In this scenario, PolicyOrg may want to search for articles based on trusted sources of information. In this case, PolicyOrg views data supplied by the government as reliable. However, when searching for content, it is not always clear whether it is derived from data coming from such a reliable source. For example, while the chart in Bob's article is based on GovStat, that information is somewhat hidden because the chart was produced by a fairly complex process performed by Alice.

Use Case 2.3   Finding flawed figures   Nick, the webmaster, discovers in his final checks before publishing Bob's article that the included chart seems off; one of the figures appears to be too high. Nick wants to find when the error occurred and who was responsible for it so that he can notify them both that there is an error and where in the production pipeline that error occurred. In this case, Alice or Bob could use that information to fix the chart. In general, the question is how to quickly identify the source of errors.

2.2.2   COMPLIANCE

Rules, regulations, policies, contracts and other sorts of guidelines provide a framework for people to do their work. Broadly, such rules describe what *should* happen in any given environment. However, practically, it is often impossible to ensure that every rule is correctly followed as work is being performed. Instead, organizations check whether a given process *complies* with the set of rules after the fact. In the data journalism case, there are particular instances where there is a need to check compliance.

Use Case 2.4   Following policy   In order to ensure that articles are of high-quality and fit the overall editorial vision of the site, NowNews has established a publication policy. Part of this publication policy is that all posts should be checked by an editor before final publication. It is up to Nick, the webmaster, to ensure that this check has taken place. NowNews also does an internal audit every year to ensure that the publication policy has been followed and to identify, in cases where it was not followed, what happened.

Use Case 2.5   Licensing   NowNews relies on multiple content providers for content pieces; this includes wire services, stock photo publishers, and governmental organizations. It tries to ensure that the content it reuses is properly attributed and that it is reused according to the correct license. However, NowNews is hit with a lawsuit claiming that they are infringing copyright of OtherNews by inappropriately using statistics retrieved from that site. To defend themselves, NowNews needs to provide the evidence (i.e., the licenses) for all data they acquired from external organizations.

2.2.3 CATALOGING

Both users and organizations spend time organizing, cataloging, and inventorying their work in order to make it more accessible and reusable. For users, this allows them to quickly find information or artifacts that they need. For organizations, such catalogs allow them to understand what their output is and who contributed to it. NowNews has these requirements as well.

Use Case 2.6 Building an index To increase both its visibility and authority, NowNews wants to provide an up-to-date catalog of all of the data sources and methods (like scripts) it uses on its website. In addition, this index provides data crunchers, such as Alice, quick access to other data sources and scripts that she might want to use for new stories.

Use Case 2.7 Acknowledgments For many of its articles, NowNews relies on the integration of multiple data sources. In order to ensure correct credit is given, NowNews wants to provide a central acknowledgments list that recognizes all the people and data sources that contribute to all the various articles and information that it publishes.

2.2.4 REPLAY

In dynamic environments, processes are often performed in ad hoc fashion or are only partially systematized: a PowerPoint document is quickly put together for the talk the next day; a budget analysis is adjusted on-site to meet the customer's need; or a design emerges from a team meeting. Given this flexibility, it is often difficult to understand what happened in particular when processes are dependent on critical people (i.e., there is lack of organizational memory). Just like a slow-motion replay in sports, it would often be useful to be able to replay these sorts of processes. Here are two examples from data journalism.

Use Case 2.8 Reproducibility When GovStat releases new statistics, the current article needs an updated figure. However, Alice has left the organization so it is not clear how the figure was created. A new data cruncher would like to reproduce what Alice did to create the updated figure.

Use Case 2.9 Publication Embargo GovStat releases its data early to a select set of organizations including NowNews, so that they can prepare articles that are timed to the release of the employment data. This data is confidential until after the embargo period has elapsed. Somehow, Bob's article on the statistics is published before that embargo. NowNews wants to determine how the confidential data got released.

2.3   A BRIEF INTRODUCTION TO EXPRESSING PROVENANCE

The above use cases can be solved in a variety of ways. Throughout the rest of this book we discuss how they can be tackled by making use of provenance, which is essentially the digital equivalent of a paper trail that describes what occurred in a system. Here, we walk through the core structures for expressing provenance, illustrating them with this scenario. The entire provenance for this scenario can be found online.[4] We use only a portion of the full provenance starting with the article produced by Bob and tracing backwards in time.

The article itself is what PROV terms an entity. We describe things like articles, data sets, charts or web pages as entities. A simple fact that we might like to express is the attribution of the article to Bob. In PROV, we represent people and organizations using the Agent construct. Using these concepts, Figure 2.3 shows an edge in a *provenance graph* describing the provenance of the employment article. In this case, its attribution.

Figure 2.3: The employment article was attributed to Bob.

This shows one view of provenance, the *responsibility* view, which describes who or what bears the responsibility for some action or thing. Attribution is a simple form of responsibility but, as we see later, PROV allows for chains of responsibility to be described. For example, while Bob is responsible for the article, the NowNews organization is as well.

While attribution is an important part of provenance, a key part of understanding how the article came to be is to understand what the article builds upon. For example, in Use Case 2.7, when creating a catalog of acknowledgments, one needs to know what data the article builds upon. Likewise, in ascertaining trust or determining information about licenses for content, it is necessary to trace back to the data or other information that Bob used to write the article. To represent the derivation of one entity from another, `prov:wasDerivedFrom` is used. Figure 2.4 expresses how `now:employment-article-v1.html` was derived from a file named `govftp:oesm11st.zip`, the data provided by GovStat.

We have now extended the provenance graph to also represent the data that the employment article relied upon. We can in turn describe how the file `govftp:oesm11st.zip` depended on other entities. Likewise, we can describe that, for instance, the report produced by PolicyOrg was derived from this employment article.

---

[4]Scenario trace: `http://www.provbook.org/provapi/documents/d000.svg`

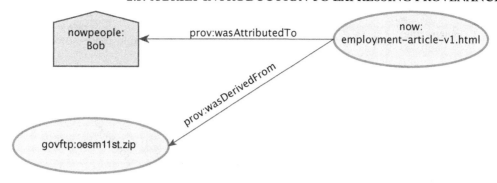

Figure 2.4: The employment article was derived from the GovStat data

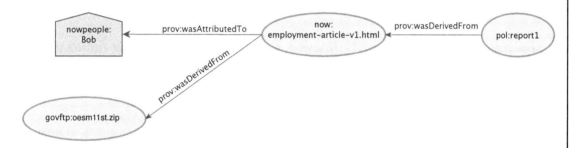

Figure 2.5: A chain of derivations from the policy report through the employment article to a data file.

This chain of derivations represents another view of provenance: the *data flow view*. It expresses the dependencies between data in a system. Thus far, we have described who is responsible for the employment article, how the article both builds upon other data and how other content depends upon the article, but we have not described *how* or by what process the employment article was created. Essentially, we would like to expand the `prov:wasDerivedFrom` relationship to incorporate how the data file was used to create the article.

In PROV, things that occurred or happened are represented by the `prov:Activity` construct. Entities are used and generated by activities. Thus, we can refine the derivation by describing how the employment article was generated by the writing activity which used the file `govftp:oesm11st.zip` as shown in Figure 2.6.

The addition of activities provides the third view of provenance, that of *process flow*. This view allows for the description of chains of operations. Introducing activities into a provenance graph provides a powerful mechanism for enriching provenance. For example, we can describe the length of time the writing process took by annotating the activity with time periods indicating

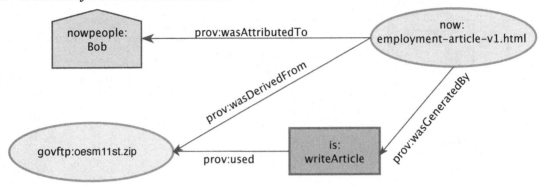

Figure 2.6: The employment article was generated by a writing activity, which used government statistics data.

its beginning and end (see Figure 2.7). Likewise, we can describe how different entities played different *roles* in the writing process. For example, we could distinguish between core and ancillary material.

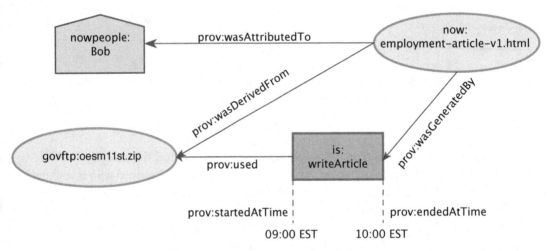

Figure 2.7: Writing began at 9 am and ended at 10 am.

Building on the three simple constructs of Entity, Activity, and Agent and their connections through data flow, process flow, and responsibility relations, we can progressively express more complex provenance. This ability to build up provenance is a key capability of PROV, as it allows one to easily get started and then refine provenance by adding more detail. Here, we expressed the provenance of Bob's article on employment, tracing it back through the writing process and to a file provided by GovStat. We can continue to further build on this, for example, by describing how Alice transformed the data into a figure. The reader is invited to check online to see the full trace of the scenario. While this is a simple provenance trace, one can begin to see how use cases such as timeliness and cataloging can be addressed by simply tracing back through the captured provenance. In the rest of the book, the expanded example will be used to show how these use cases can be dealt with in a more complete fashion.

2.4 SUMMARY

In this chapter, we presented a data journalism scenario and illustrated four categories of use cases: quality assessment, compliance, cataloging and replay. We then showed how part of the scenario could be captured using the basic building blocks of PROV and provided an intuition as to how some of the use cases would be addressed. In the next chapter, we expand on these core building blocks describing the PROV ontology in more detail.

CHAPTER 3

The PROV Ontology

Chapter 2 introduced a series of provenance use cases (Section 2.2) and intuitively described the constituents of a provenance vocabulary (Section 2.3). From Section 1.4, we recall that PROV is structured around a conceptual data model (PROV-DM [33]) and various serializations. PROV-O [27] is an OWL2 ontology allowing a mapping of PROV to RDF, which itself offers serializations such as TURTLE and RDF/XML.

This chapter aims to provide a compact, simplified presentation of PROV-O, which can be used as a quick reference to the various classes and properties used in this book. Definitions are copied verbatim from PROV-DM [33]. For a detailed presentation, the reader is invited to consult normative presentations of PROV [27, 33].

3.1 OVERVIEW

There are several ways of structuring the presentation of the PROV ontology. Our presentation in this chapter is organized according to three different views of provenance, which users can adopt, according to their preference and their application needs. The three views are the *data flow view*, the *process flow view*, and the *responsibility view*, which we introduce below. Furthermore, PROV can be divided into a core, forming the essence of provenance, and a set of extended constructs, catering to more advanced uses of provenance. Figure 3.1 displays the three views of provenance and the associated core classes and properties.

The core of PROV is formed[2] of three classes and seven properties. They are organized as follows.

1. The *data flow view* is concerned with the flow of information inside computer systems or the transformation of things in physical or imaginary worlds. Entities (class prov:Entity) are digital artifacts or arbitrary things we want to describe the provenance of. The transformation and the flow of these entities is what we refer to as a *Derivation*, which is encoded by the property prov:wasDerivedFrom.

2. At times, it is useful to provide more information about derivations, by listing the processes that took place and all associated timing information. The data flow view can then be refined by the *process flow view* enumerating the activities that occurred, as well as their start and end times. Activities (class prov:Activity) take entities as input (it is the notion of *Usage*,

---

[2]Namespace prefixes and URIs can be found in Table 1.3.

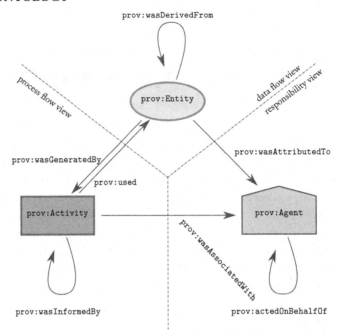

Figure 3.1: Three different views of the core of PROV-O. The figure adopts the PROV layout conventions:[1] an entity is represented by a yellow ellipsis, an activity by a blue rectangle, and an agent by an orange pentagon house. We note here that the diagram is a "class diagram" illustrating the classes that occur as domain and range of properties.

described by the property prov:used) and output new entities (it is the notion of *Generation* described by the property prov:wasGeneratedBy). Furthermore, data may have flown from one activity to another, which is captured by the concept of *Communication*, expressed by prov:wasInformedBy.

3. Provenance is also about assigning responsibility for what happened in a system, i.e. the *responsibility view*. Agent (class prov:Agent) is the class of things found in the range of three properties. Agents may be responsible for (1) the existence of entities: that is what we call *Attribution*, which is expressed by the property prov:wasAttributedTo; (2) for past activities: that is an *Association*, which is expressed by prov:wasAssociatedWith, or (3) for other agents: that constitutes a *Delegation*, encoded by prov:actedOnBehalfOf.

It should be noted that properties have a verbal form in the past to remind the reader that provenance provides descriptions about past executions. This is an important aspect of provenance. Provenance is not a programming language, describing what may happen when the program is executed. Instead, *it is a description of what has happened*.

3.2   QUALIFIED RELATION PATTERNS

The core of PROV is intentionally kept simple (three classes and seven properties) to ease the creation of RDF triples, and therefore to promote adoption. In addition, PROV provides further mechanisms to refine provenance, allowing levels of detail to support some advanced application requirements.

Sometimes, a binary relation is not sufficient to describe a situation: for example, we may want to indicate the time at which an entity was generated by an activity, or we may want to specify the activity for which a delegation of responsibility took place. PROV adopts the *Qualified Relation pattern* [10] as a systematic mechanism to reify binary relations, and to allow further information to be linked up; in contrast, binary relations, for which the qualified relation pattern has not been applied, are referred to as *unqualified relations*.

Figure 3.2 illustrates the qualified relation pattern. Let us consider a binary relation between two resources r_1, r_2, expressed by triple $(r_2\ \mathtt{prov:}XXX\ r_1)$ where *XXX* can be one of used, wasGeneratedBy, wasDerivedFrom, etc. The source of this relation, r_2, and the destination, r_1, are respectively referred to as influencee and influencer. The qualified relation pattern introduces a new resource x, which is an instance of the class representing the qualified relation; x is a reified representation of an unqualified relation, which in turn can be annotated with further descriptions. The influencee is then linked to resource x, with a property named qualified*XXX*, where *XXX* can be Usage, Generation, Derivation, etc. The new resource x is of type *XXX*. The new resource is then linked to the influencer, with the property prov:influencer. A property chain that links the influencee to the influencer, via the qualified relation instance, is intended to carry the same meaning as an unqualified relation directly linking an influencee to an influencer.

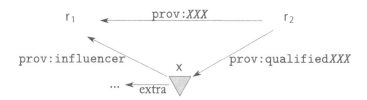

Figure 3.2: The Qualified Relation Pattern for an unqualified relation linking an influencee r_2 to an influencer r_1.

Figures 3.4, 3.6, 3.7, 3.8, 3.9, 3.10, 3.11, 3.13, 3.14, and 3.15 illustrate the qualified relation pattern for the various PROV properties it applies to. Table 3.1 shows how each property for an unqualified relation maps to a class for a qualified relation; the first part of the table lists core PROV properties, whereas the second focuses on extended properties.

Table 3.1: Mapping of properties for unqualified relations to classes for qualified relations

Property for Unqualified Relation	Class for Qualified Relation
prov:wasDerivedFrom	prov:Derivation
prov:wasGeneratedBy	prov:Generation
prov:used	prov:Usage
prov:wasInformedBy	prov:Communication
prov:wasAttributedTo	prov:Attribution
prov:wasAssociatedWith	prov:Association
prov:actedOnBehalfOf	prov:Delegation
prov:wasStartedBy	prov:Start
prov:wasEndedBy	prov:End
prov:wasInvalidatedBy	prov:Invalidation
prov:wasInfluencedBy	prov:Influence

3.3 DATA FLOW VIEW

The data flow view focuses on the flow of information and the transformation of things in systems. Information items and things are termed entities, whereas flows and transformations are termed derivations.

3.3.1 ENTITY

An entity is a physical, digital, conceptual, or other kind of thing with some fixed aspects; entities may be real or imaginary.

Example   An article, JPG file, and data set are all entities.

:e a Entity.

Figure 3.3: Entity.

Class	Parent
prov:Entity	—

3.3.2 DERIVATION

A derivation is a transformation of one entity into another, an update of an entity resulting in a new one, or the construction of a new entity based on a pre-existing entity.

Example   The JPG file (entity :e2), containing a plot, was derived from the data set (entity :e1).

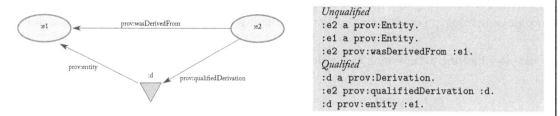

```
Unqualified
:e2 a prov:Entity.
:e1 a prov:Entity.
:e2 prov:wasDerivedFrom :e1.
Qualified
:d a prov:Derivation.
:e2 prov:qualifiedDerivation :d.
:d prov:entity :e1.
```

Figure 3.4: Illustration of Qualified and Unqualified Derivation.

Class	Parent	
prov:Derivation	prov:Influence	
Property	Domain	Range
prov:wasDerivedFrom	prov:Entity	prov:Entity
prov:qualifiedDerivation	prov:Entity	prov:Derivation
prov:entity	prov:Influence	prov:Entity

3.3.3   REVISION

A revision is a derivation for which the resulting entity is a revised version of some original.

Example   As a new data set was published by the government, the article was adapted to rely on the latest data set: the new article is a revision of the preceding article.

3.3.4   QUOTATION

A quotation is the repeat of (some or all of) an entity, such as text or image, by someone who may or may not be its original author.

Example   Bob's article included a paragraph that was quoted from a third-party document.

3.3.5   PRIMARY SOURCE

A primary source for a topic refers to something produced by some agent with direct experience and knowledge about the topic, at the time of the topic's study, without benefit from hindsight.

Example   The primary source for the plot was the governmental data set.

3.4   PROCESS FLOW VIEW

The process flow view refines the data flow view by introducing activities, their inter-relations with entities, and their respective times.

3.4.1   ACTIVITY

An activity is something that occurs over a period of time and acts upon or with entities; it may include consuming, processing, transforming, modifying, relocating, using, or generating entities.

Example   Plotting data, writing an article, and compiling a survey are activities.

```
:a a Activity.
```

Figure 3.5: Activity.

Class	Parent	
prov:Activity	—	
Property	Domain	Range
prov:startedAtTime	prov:Activity	xsd:dateTime
prov:endedAtTime	prov:Activity	xsd:dateTime

3.4.2   GENERATION

Generation is the completion of production of a new entity by an activity. This entity did not exist before generation and becomes available for usage after generation.

Example   The JPG file (entity : e) was generated by the plotting execution (activity : a) at 12:01:01 on April 1, 2012.

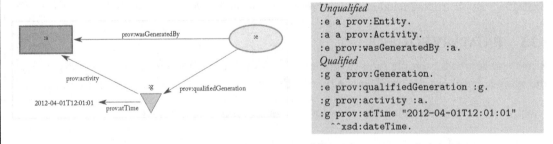

```
Unqualified
:e a prov:Entity.
:a a prov:Activity.
:e prov:wasGeneratedBy :a.
Qualified
:g a prov:Generation.
:e prov:qualifiedGeneration :g.
:g prov:activity :a.
:g prov:atTime "2012-04-01T12:01:01"
  ^^xsd:dateTime.
```

Figure 3.6: Illustration of Qualified and Unqualified Generation.

Class	Parent	
prov:Generation	prov:ActivityInfluence, prov:InstatenousEvent	
Property	Domain	Range
prov:wasGeneratedBy	prov:Entity	prov:Activity
prov:qualifiedGeneration	prov:Entity	prov:Generation
prov:activity	prov:ActivityInfluence	prov:Activity
prov:atTime	prov:InstantaneousEvent	xsd:dateTime

3.4.3 USAGE

Usage is the beginning of utilizing an entity by an activity. Before usage, the activity had not begun to utilize this entity and could not have been affected by the entity.

Example   The uncompressing activity (:a) used the zip file (entity :e) at 12:01:01 on April 1, 2012.

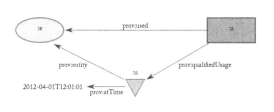
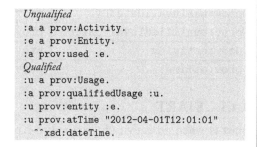

Unqualified
```
:a a prov:Activity.
:e a prov:Entity.
:a prov:used :e.
```
Qualified
```
:u a prov:Usage.
:a prov:qualifiedUsage :u.
:u prov:entity :e.
:u prov:atTime "2012-04-01T12:01:01"
  ^^xsd:dateTime.
```

Figure 3.7: Illustration of Qualified and Unqualified Usage.

Class	Parent	
prov:Usage	prov:EntityInfluence, prov:InstantaneousEvent	
Property	Domain	Range
prov:used	prov:Activity	prov:Entity
prov:qualifiedUsage	prov:Activity	prov:Usage
prov:entity	prov:EntityInfluence	prov:Entity
prov:atTime	prov:InstantaneousEvent	xsd:dateTime

3.4.4 INVALIDATION

Invalidation is the start of the destruction, cessation, or expiry of an existing entity by an activity. The entity is no longer available for use (or further invalidation) after invalidation. Any generation or usage of an entity precedes its invalidation.

Example   A file (entity :e) was deleted from the file system. Hence, file :e was invalidated by this deletion activity.

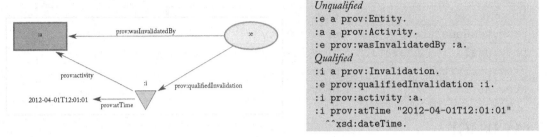

```
Unqualified
:e a prov:Entity.
:a a prov:Activity.
:e prov:wasInvalidatedBy :a.
Qualified
:i a prov:Invalidation.
:e prov:qualifiedInvalidation :i.
:i prov:activity :a.
:i prov:atTime "2012-04-01T12:01:01"
   ^^xsd:dateTime.
```

Figure 3.8: Illustration of Qualified and Unqualified Invalidation.

Class	Parent	
prov:Invalidation	prov:ActivityInfluence, prov:InstantaneousEvent	
Property	Domain	Range
prov:wasInvalidatedBy	prov:Entity	prov:Activity
prov:qualifiedInvalidation	prov:Entity	prov:Invalidation
prov:activity	prov:ActivityInfluence	prov:Activity
prov:atTime	prov:InstantaneousEvent	xsd:dateTime

3.4.5   START

Start is when an activity is deemed to have been started by an entity, known as a trigger. The activity did not exist before its start. Any usage, generation, or invalidation involving an activity follows the activity's start. A start may refer to a trigger entity that set off the activity, or to an activity, known as a starter, that generated the trigger.

Example   Upon receipt of the signed contract (entity :e), the activity of writing the article (activity :a) was started; the contract was the outcome of a long negotiation activity (activity :a1).

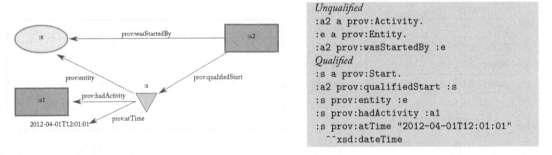

```
Unqualified
:a2 a prov:Activity.
:e a prov:Entity.
:a2 prov:wasStartedBy :e
Qualified
:s a prov:Start.
:a2 prov:qualifiedStart :s
:s prov:entity :e
:s prov:hadActivity :a1
:s prov:atTime "2012-04-01T12:01:01"
   ^^xsd:dateTime
```

Figure 3.9: Illustration of Qualified and Unqualified Start

Class	Parent	
prov:Start	prov:EntityInfluence, prov:InstantaneousEvent	

Property	Domain	Range
prov:wasStartedBy	prov:Activity	prov:Entity
prov:qualifiedStart	prov:Activity	prov:Start
prov:entity	prov:EntityInfluence	prov:Entity
prov:hadActivity	prov:Influence	prov:Activity
prov:atTime	prov:InstantaneousEvent	xsd:dateTime

3.4.6   END

End is when an activity is deemed to have been ended by an entity, known as a trigger. The activity did not exist before its end. Any usage, generation, or invalidation involving an activity follows the activity's end. An end may refer to a trigger entity that set off the activity, or to an activity, known as an ender, that generated the trigger.

Example   As soon as the article (entity :e) was published (activity :a1), the writing activity (activity :a2) was deemed to have ended.

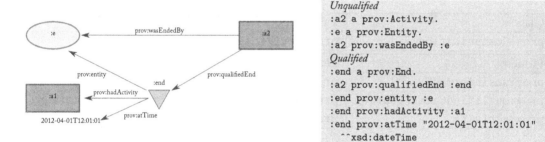

Figure 3.10: Illustration of Qualified and Unqualified End

Class	Parent	
prov:End	prov:EntityInfluence, prov:InstantaneousEvent	

Property	Domain	Range
prov:wasEndedBy	prov:Activity	prov:Entity
prov:qualifiedEnd	prov:Activity	prov:End
prov:entity	prov:EntityInfluence	prov:Entity
prov:hadActivity	prov:Influence	prov:Activity
prov:atTime	prov:InstantaneousEvent	xsd:dateTime

3.4.7 COMMUNICATION

Communication is the exchange of some unspecified entity by two activities, with one activity using some entity generated by the other.

Example   The task of writing an article (activity :a2) was informed by the task assignment (activity :a2); a brief was provided, but is not explictly modeled.

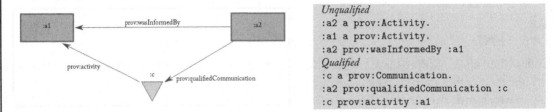

Figure 3.11: Illustration of Qualified and Unqualified Communication.

Class	Parent	
prov:Communication	prov:ActivityInfluence	
Property	Domain	Range
prov:wasInformedBy	prov:Activity	prov:Activity
prov:qualifiedCommunication	prov:Activity	prov:Communication
prov:activity	prov:ActivityInfluence	prov:Activity

3.5 RESPONSIBILITY VIEW

The responsibility view is concerned with the assignment of responsibility to agents for what happened in a system.

3.5.1 AGENT

An agent is something that bears some form of responsibility for an activity taking place, for the existence of an entity, or for another agent's activity.

Example   Bob and Alice are agents; Bob delegates responsibility to Alice, who is in charge of processing a data set.

:ag a Agent.

Figure 3.12: Agent.

Class	Parent
prov:Agent	—

3.5.2   ATTRIBUTION

Attribution is the ascribing of an entity to an agent.

Example   The article was attributed to agent Bob.

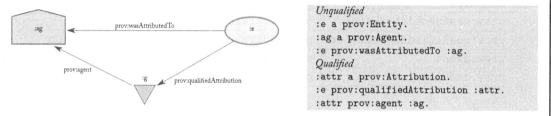

Figure 3.13: Illustration of Qualified and Unqualified Attribution.

Class	Parent	
prov:Attribution	prov:AgentInfluence	
Property	Domain	Range
prov:wasAttributedTo	prov:Entity	prov:Agent
prov:qualifiedAttribution	prov:Entity	prov:Attribution
prov:agent	prov:AgentInfluence	prov:Agent

3.5.3   ASSOCIATION

An association is an assignment of responsibility to an agent for an activity, indicating that the agent had a role in the activity. It further allows for a plan to be specified, which is the plan intended by the agent to achieve some goals in the context of this activity.

Example   The agent responsible for the publishing activity is Nick; he proceeded according to the publication policy (a Plan).

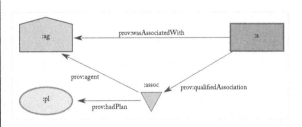

```
Unqualified
:a a prov:Activity.
:ag a prov:Agent.
:a prov:wasAssociatedWith :ag.
Qualified
:assoc a prov:Association.
:pl a prov:Plan.
:a prov:qualifiedAssociation :assoc.
:assoc prov:agent :ag.
:assoc prov:hadPlan :pl.
```

Figure 3.14: Illustration of Qualified and Unqualified Association.

Class	Parent	
prov:Association	prov:AgentInfluence	
prov:Plan	prov:Entity	
Property	Domain	Range
prov:wasAssociatedWith	prov:Activity	prov:Agent
prov:qualifiedAssociation	prov:Activity	prov:Association
prov:agent	prov:AgentInfluence	prov:Agent
prov:hadPlan	prov:Association	prov:Plan

3.5.4 DELEGATION

Delegation is the assignment of authority and responsibility to an agent (by itself or by another agent) to carry out a specific activity as a delegate or representative, while the agent it acts on behalf of retains some responsibility for the outcome of the delegated work.

Example   Agent Alice, analyzing the data set, acted on behalf of agent Bob.

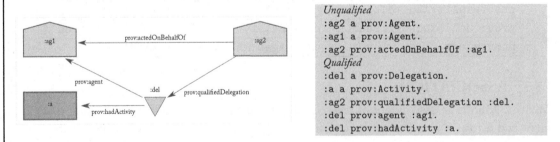

```
Unqualified
:ag2 a prov:Agent.
:ag1 a prov:Agent.
:ag2 prov:actedOnBehalfOf :ag1.
Qualified
:del a prov:Delegation.
:a a prov:Activity.
:ag2 prov:qualifiedDelegation :del.
:del prov:agent :ag1.
:del prov:hadActivity :a.
```

Figure 3.15: Illustration of Qualified and Unqualified Delegation.

Class	Parent	
prov:Delegation	prov:AgentInfluence	
Property	Domain	Range
prov:actedOnBehalfOf	prov:Agent	prov:Agent
prov:qualifiedDelegation	prov:Agent	prov:Delegation
prov:agent	prov:AgentInfluence	prov:Agent
prov:hadActivity	prov:Delegation	prov:Activity

3.6   ALTERNATES VIEW

The alternates view is a more specialist notion of provenance, which is concerned with linking entities that present different facets of the same thing. An article may be known by a URI irrespective of its various versions, which may also have their own specific URIs. The article in general and its versioned copies may be related as follows. We note that PROV-O does not define the Qualified Relation pattern for properties defined in this section.

3.6.1   SPECIALIZATION

An entity that is a specialization of another shares all aspects of the latter, and additionally presents more specific aspects of the same thing as the latter. In particular, the lifetime of the entity being specialized contains that of any specialization.

Example   The second version of the article on employment data by NowNews is a specialization of the article on employment data by NowNews (irrespective of its version).

:e2 prov:specializationOf :e1.

Figure 3.16: Specialization.

Property	Domain	Range
prov:specializationOf	prov:Entity	prov:Entity

3.6.2   ALTERNATE

Two alternate entities present aspects of the same thing. These aspects may be the same or different, and the alternate entities may or may not overlap in time.

Example   The second article on employment data by NowNews is an alternate of the first version of this article (and vice-versa).

:e2 prov:alternateOf :e1.

Figure 3.17: Alternate.

Property	Domain	Range
prov:alternateOf	prov:Entity	prov:Entity

3.7   BUNDLES

A bundle (class prov:Bundle) is a named set of provenance descriptions, and is itself an entity, allowing provenance of provenance to be expressed.

Example   Alice packaged up the provenance of the JPG file as a bundle and embedded it in the file.

PROV does not provide a specific construct to support bundles, but instead relies on the serialization format to express these. An approach, adopted in the following snippet, is to use the TRIG notation to represent a named graph. A bundle is also an entity, and its provenance can be described.

```
:b1  {
    :f a prov:Entity, schema:Imageobject.
    :f prov:wasGeneratedBy :gnuplot.
}

{
    :b1 a prov:Entity, a prov:Bundle.
    :b1 prov:wasAttributedTo nowpeople:Alice
}
```

3.8   MISCELLANEOUS

3.8.1   COLLECTION AND MEMBERSHIP

A collection (class prov:Collection) is an entity that provides a structure to some constituents that must themselves be entities. These constituents are said to be members of the collections. An empty collection is a collection without members.

Example   NowNews article (entity :e) was one of the many articles included in the PolicyOrg compilation (collection :c).

```
:c a prov:Collection.
:e a prov:Entity.
:c prov:hadMember :e.
```

Figure 3.18: Illustration of Membership Property.

Class	Parent	
prov:Collection	prov:Entity	

Property	Domain	Range
prov:hadMember	prov:Collection	prov:Entity

3.8.2 REFINED DERIVATION

In some cases, it is useful to make explicit the full path underpinning a derivation; this includes an activity, a generation, and a usage. They are explicitly linked using three properties (displayed in black in Figure 3.19). Properties in gray were respectively defined in Figures 3.7 and 3.6.

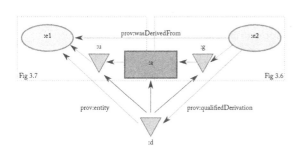

Unqualified
```
:e2 a prov:Entity.
:e1 a prov:Entity.
:e2 prov:wasDerivedFrom :e1.
```
Qualified
```
:d a prov:Derivation.
:e2 prov:qualifiedDerivation :d.
:d prov:entity :e1.
:d prov:hadActivity :a.
:d prov:hadGeneration :g.
:e2 prov:hadQualifiedGeneration :g.
:d prov:hadUsage :u.
:a prov:hadQualifiedUsage :u.
```

Figure 3.19: Refined Qualified Derivation.

Class	Parent	
prov:Derivation	prov:EntityInfluence	

Property	Domain	Range
prov:hadActivity	prov:Influence	prov:Activity
prov:hadGeneration	prov:Derivation	prov:Generation
prov:hadUsage	prov:Derivation	prov:Usage

3.8.3   FURTHER PROPERTIES

A value is a constant such as a string, number, time, qualified name, IRI, and encoded binary data, whose interpretation is outside the scope of PROV.

Example   The zip utility was called with the "-9" parameter to optimize compression: this parameter is an entity that has the string "-9" as a value.

```
:param a prov:Entity;
   prov:value "-9".
```

A location can be an identifiable geographic place (ISO 19112), but it can also be a non-geographic place such as a directory, row, or column.

Example   NowNews offices are located in New York.

```
:e a prov:Entity, ex:Office;
   prov:atLocation <http://dbpedia.org/resource/New_York>.
```

A role is the function of an entity or agent with respect to an activity, in the context of a usage, generation, invalidation, association, start, and end.

Example   Nick was associated with the publication activity in his role of publisher.

```
is:pub prov:qualifiedAssociation [ a prov:Association ;
                        prov:agent nowpeople:Nick;
                        prov:hadRole prov:Publisher ].
```

Class	Parent	
prov:Location	—	
prov:Role	—	
Property	Domain	Range
prov:value	prov:Entity	rdfs:Literal
prov:atLocation	owl:Thing	prov:Location
prov:hadRole	prov:Influence	prov:Role

3.9   ONTOLOGY STRUCTURE

PROV also encompasses several organizing elements to structure the ontology. These include the upper-level classes prov:Influence and prov:InstantaneousEvent. Influence is defined as the capacity of an entity, activity, or agent to have an effect on the character, development, or behavior of another by means of usage, start, end, generation, invalidation, communication, derivation, attribution, association, or delegation. The PROV data model is further implicitly based on a notion of instantaneous events that mark transitions in the world (for further description see Section 5.2.1).

The classes and properties presented in this chapter are further organized according to the hierarchies of Figure 3.20. All qualified classes in the domain of prov:atTime are subclasses of prov:InstantaneousEvent. All qualified classes have prov:Influence as a superclass. This hierarchy is also reflected in properties: prov:qualifiedInfluence is a super-property of all "qualifiedXXX", and prov:wasInfluencedBy is a super-property of all unqualified relations. Furthermore, prov:Influence is subclassed into prov:EntityInfluence, prov:ActivityInfluence, and prov:AgentInfluence when influencers are respectively prov:Entity, prov:Activity, and prov:Agent.

Further subclasses are introduced. Common types of agents are identified by means of the classes prov:SoftwareAgent, prov:Organization, and prov:Person. These subclasses are a good illustration of how PROV is intended to be extended. Developers are invited to extend the core concepts of PROV, including prov:Entity, prov:Activity, and prov:Agent, as well as the classes shown in Figure 3.20 with a blue-bold border. By restricting extensions to these classes and associated properties, they insure interoperability and conversion to the various representations of PROV.

3.10 SUMMARY

The organization of PROV-O in different views of provenance allows for an incremental approach to asserting provenance. First, the designer can select simple concepts to express provenance assertions in an easy manner. This approach is referred to as "scruffy" because provenance assertions may not have all the necessary precision, or may not express past executions in minute details. However, scruffy assertions can be refined by adding a further process flow view to refine the data flow view; by expressing qualified patterns to provide further information to unqualified relations; or by expressing alternates view to link up related resources. Building on this vocabulary, the next chapter proposes a set of recipes to help the designer to construct extensive provenance descriptions.

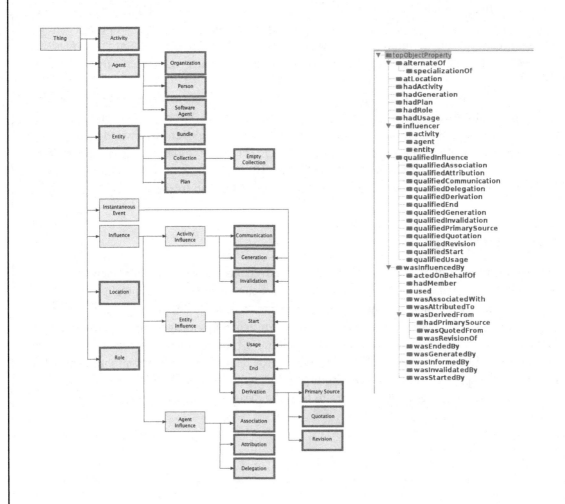

Figure 3.20: Classes and Properties Hierarchy in the PROV Ontology. Blue, bold bordered boxes denote extension point classes.

CHAPTER 4

Provenance Recipes

PROV provides structures and terms for modeling provenance. However, given the variety of applications, the PROV specifications themselves do not provide concrete modeling and usage guidance. In this chapter, we provide a series of recipes for both expressing and integrating provenance into applications. This guidance is derived from our experience producing multiple provenance-enabled applications [28]. These recipes do not cover all questions concerning provenance in applications; instead, just like basic recipes in cooking, they should provide the foundations for creating one's own provenance-enabled application.

We organize the recipes into guidance for modeling provenance, organizing provenance information, and collecting provenance, and, then address some common pitfalls. Each recipe follows this pattern:

- Question — the question we are trying to address with the recipe;

- Context — the context where the recipe applies;

- Solution — a set of instructions for answering the question, usually step-by-step;

- Scenario – an illustration of where this recipe could be used in the data journalism scenario; and

- Discussion — a discussion of issues related to the recipe and pointers to related information.

4.1 MODELING

This section defines recipes related to modeling the provenance of an application or environment.

4.1.1 ITERATIVE MODELING

Question: How does one start to integrate PROV into an application?

Context: An application currently does not produce provenance or produces it according to its own schema. If a developer wants to integrate PROV into the application, what is a good way to proceed?

Solution: We suggest a four-step approach to integrating PROV:

1. Identify - The first step is to identify the core data elements (or types) within the application. Data items (or instances) of these elements are what one will later want to determine the provenance of.

2. Model - Describe how these data elements are connected together via PROV constructs. We suggest starting with a high-level provenance description at first and then further refining the description. See Section 4.1.3 for more details.

3. Instrument - Modify the application to export PROV according to the defined model.

4. Query and Refine - Once the application exports provenance, run a small trial of the application, and query the provenance according to provenance use cases. Based on this testing, refine the model by introducing additional provenance information or identifying new data elements.

Scenario: A good example from the data journalism scenario, where this approach can be applied, is the creation of the chart by Alice as described in Section 2.1.3. As an application designer, we would identify that data files, such as the resulting chart, are key elements that we want to get the provenance of. We would then model how the usage and generation of these data files happen with respect to command line programs. Finally, we could instrument or change Alice's environment. We could subsequently refine the provenance to also model more explicitly types of data (e.g., CSV and RDF) and the program itself (e.g., gnuplot).

Discussion: A key notion in this approach is to iteratively refine provenance. Applications can always be described in more detail. For instance, a user submitting some information to a system could be documented by attributing that information to that user; this can be refined by adding an activity, the running process, its timing, configuration, and logging details preceding the submission; this itself can be refined by describing the flow of information in the application, including access control checking, allocation of resources, etc.; ultimately, one could describe the individual instructions executed by the processor. There is a cost-benefit trade-off in the level of detail that one uses to describe provenance. By iteratively adding more details to provenance, applications can provide useful information right away without trying to model every minute action. In terms of instrumenting applications, we discuss more approaches in Section 4.3.

4.1.2   IDENTIFY, IDENTIFY, IDENTIFY!

Question: How does one ensure that provenance for a data item can be retrieved?

Context: A user of provenance information wants to be able to retrieve the provenance of a data item. It is useful to provide users with appropriate handles to data within provenance and also to be clear about what these handles refer to.

Solution: The suggested approach is to create an identifier scheme within the application such that all data items get assigned a unique identifier, that is, the data items are treated as entities.

1. Create an identifier scheme. Such a scheme provides a pattern for the generation of identifiers for data items during the execution of an application. We suggest using HTTP URLs that are dereferenceable. In Section 6.3, we describe the identifier scheme for this book.

2. For each data item, specify the fixed aspects of that item. Fixed aspects are those properties of the data item that will not change, for example, a version number, the identifier itself, the location of the data item, its type, or its content.

3. Associate each data item with a type and/or a specialization hierarchy. Because each data item is produced per execution of an application, embedding it within a hierarchical structure, for example by giving it a type or associating it with a more general entity, helps in writing queries to locate the item.

4. For every data item, generate an identifier according to the identifier scheme and use it within the provenance.

Scenario: NowNews wants to set up an identifier scheme for its content that ensures that each version of an article has its own versioned URL. These would then be connected to a page that represents the article as a whole. Here the fixed aspect would be the version number.

Discussion: Having an identifier is central to connecting provenance graphs as it allows one to refer to the various parts of the graph (e.g., entities, activities, agents). The use of HTTP URLs for provenance enables them to be subsequently exposed as Linked Data and, thus, to be queried in combination with other data sets. A central issue is the specification of fixed aspects. Section 4.1.4 gives more details about fixed aspects with respect to versioning. A good example of identification is web-based version control systems such as Github as they provide precise fixed-aspects (i.e. a hash) and are dereferenceable.

4.1.3   FROM DATA FLOW TO ACTIVITIES

Question: Which provenance view should be modeled first?

Context: As detailed in Chapter 3, PROV supports three views of provenance: data flow, process flow, and responsibility. When beginning the modeling of an application, it is often helpful to know which view to model first.

Solution: We suggest focusing on data first, since entities tend to be easily identifiable. This solution builds on Recipe 4.1.2.

1. Express data flow by connecting entities using `prov:wasDerivedFrom` properties.

2. Express responsibility. Once data flow has been expressed, connect the agents that are responsible for each data item using `prov:wasAttributedTo` properties. At this stage, one can also express which agents acted on behalf of other agents.

3. Refine data flow and responsibility by introducing activities. We can expand the derivation links to describe how the derivation occurred using the Refined Qualified Derivation (see Section 3.8.2).

Depending on the circumstances, modelers may want to begin by modeling responsibility first, for example, if attribution is critical for a given use case.

Scenario: This approach was followed in Section 2.3 when describing the provenance of the employment article.

Discussion: We note that this approach builds and refines the provenance graph. Maintaining linkages within the graph is important as it ensures that provenance can be traced back. Thus, whatever view one starts with, one should make sure to build these links. Beginning with data flow also ensures that the common pitfall *Activity but no Derivation* (Recipe 4.4.1) is less likely to occur.

4.1.4 PLAN FOR REVISIONS

Question: How does one express revisions to a resource or document using PROV?

Context: An important capability of provenance is to express revisions. However, for a provenance consumer, it may be difficult to find all the revisions of a single document.

Solution: We recommend connecting all versions of a resource to a single general resource using `prov:specializationOf`. For example, there is one employment article but multiple versions of it. These specific versions are specializations of the employment article in general.

1. For a resource, create an identifier to denote it in general, irrespective of its actual version, for example, the employment article.

2. For each version of the resource, create a fresh identifier for that resource, for example, the article as originally written and after an update.

3. Relate each version to the previous one using `prov:wasRevisionOf`.

4. Relate each version to the resource in general using `prov:specializationOf`.

Scenario: For the employment article and its multiple versions, applying this recipe would result in the following PROV assertions:

```
now:employment-article a prov:Entity.
now:employment-article-v1.html a prov:Entity;
                        prov:specializationOf  now:employment-article.
now:employment-article-v2.html a prov:Entity;
                        prov:specializationOf  now:employment-article.

now:employment-article-v2.html
                        prov:wasRevisionOf now:employment-article-v1.html.
```

Discussion: This recipe demonstrates the usefulness of specialization hierarchies that allow for provenance to be described at multiple levels of abstraction. For those familiar with version control systems, one can see the resource as the file itself and each of the revisions as corresponding revisions of the file. It is fairly straightforward to see how more complex revision control structures such as branching, merging and copying can be represented using PROV. A crucial benefit is that the dependencies between versions are explicit and do not rely on version numbers.

4.1.5   MODELING UPDATE AND OTHER DESTRUCTIVE ACTIVITIES

Question: How does one model the update of a entity to another?

Context: A file or other entity is updated. After the update occurs, the original file will no longer be available. This may occur in any destructive action. What is an appropriate way to model this?

Solution: There are five steps to modeling updates or other (destructive) activities. Given an updated entity u and the original entity, o, do the following:

1. Express a `prov:wasDerivedFrom` edge from u to o.

2. Model the activity itself.

3. Express that the activity used o.

4. Express that o is no longer available using a `prov:wasInvalidatedBy` edge between the activity and o.

5. Link the updated entity, u, to the activity using the `prov:wasGeneratedBy` edge.

Scenario: For example, applying this approach to the scenario, we can model that a file was updated to be an approved file as follows (essentially, the approved file replaced the old file):

```
is:f11 prov:wasDerivedFrom is:f10.
is:check9 a bk:Approval, prov:Activity.
is:check9 prov:used is:f10.
is:f10 prov:wasInvalidatedBy  is:check9.
is:f11 prov:wasGeneratedBy is:check9.
```

Discussion: We note that this approach of introducing an activity and linking it to the entity being consumed or deleted can be applied to any destructive activity. Also, we note there are several activities defined by PROV from the Dublin Core mapping [11] that may be useful in modeling provenance such as prov:Publish, prov:Replace, or prov:Modify. Using such classes encourages interoperability between systems. Finally, an activity is necessary in this recipe because an entity can only be invalidated by an activity.

4.1.6   MODELING MESSAGE PASSING

Question: How does one model the exchange of messages between different systems?

Context: Distributed systems, including the Web, heavily rely on the exchange of messages between systems. Think for, instance, of a remote procedure call, an HTTP request, or even a pipe. PROV provides the prov:wasInformedBy edge to model communication, but that does not provide the ability to describe the content of messages themselves or the actual activity of sending or receiving the message. In some cases, these parts of the communication may be useful in particular provenance use cases (for example, when using provenance information to debug an application).

Solution: Let us say that we have two activities, a1 and a2, communicating with each other, where a2 receives a message, msg, from a1 .

1. Begin by modeling the exchange of a message (msg) between the two activities using prov:wasGeneratedBy and prov:used edges:

```
:msg prov:wasGeneratedBy :a1.
:a2 prov:used :msg.
```

2. Introduce a sent message and received message connected by a derivation. These are specializations of the original message.

```
...
:sentMsg prov:wasGeneratedBy :a1.
:a2 prov:used :recvMsg.
:recvMsg prov:wasDerivedFrom :sentMsg.

:recvMsg prov:specializationOf :msg.
:sentMsg prov:specializationOf :msg.
```

Now, we can express information about the differences between what is sent and received. For example, data may be corrupted or modified during transmission and thus the sent and received message may differ.

3. We can further refine this to introduce an activity :exchange describing the actual exchange of these two messages. We also can use the usage and generation events associated with that activity to describe the specific sending and receiving events. By introducing these elements, we can add additional information about the process of exchanging a message such as how long it took..

```
:recvMsg prov:wasDerivedFrom :sentMsg.
        prov:qualifiedDerivation [
            a prov:Derivation.
            prov:entity :sentMsg.
            prov:hadActivity :exchange.
            prov:hadUsage      :sending;
            prov:hadGeneration :receiving;
        ]
```

Scenario: This recipe could be used to model the downloading of information from the government FTP site to NowNews.

Discussion: By using these steps, one can progressively model more details of the message exchange process. We note that this recipe uses the refined derivation introduced in Section 3.8.2. In [18], we describe such patterns in more detail and rules for expansion and contraction of communication as it pertains to message passing.

4.1.7   MODELING PARAMETERS

Question: How does one model parameters of application programs?

Context: Command line programs are widely used in many computational pipelines. These programs often take parameters to specify their behavior. For example, the command sort -r; provides very different output than sort.[1] Thus, capturing parameters is an important part of provenance.

Solution: Here are five steps for modeling parameters:

1. Model each parameter as an entity.

2. Connect the value of the parameter to the entity using prov:value.

3. Connect the entity to the activity using a prov:used edge.

4. Qualify the used edge.

5. Identify the role of the parameter (e.g., -r) using prov:hadRole, and adopting an appropriate namespace for the various roles.

   Here is an example describing sort -r ;.

```
:param a prov:Entity.
:param prov:value ";".
:sort a prov:Activity.
:sort prov:used :param.
:u a prov:Usage.
:sort prov:qualifiedUsage :u.
:u prov:entity :param.
:u prov:hadRole ex:dashR.
```

[1]sort -r reverses the order of the sorted results.

Scenario: This could be used in modeling the provenance of Alice's data crunching, which uses a variety of command line tools.

Discussion: This approach can also be used in modeling parameters for function or method calls. The introduction of roles using the qualified pattern is a common practice for distinguishing the role various entities play with respect to an activity. We also note that because the parameters are modeled as entities their provenance can also be described.

4.1.8   INTRODUCE THE ENVIRONMENT

Question: How do we model a system's interaction with its environment?

Context: As we have seen, provenance can be modeled at increasing levels of detail. When modeling provenance, it is often the case that a system has interactions with its environment, but there is no desire to model details of the environment itself or every interaction by the system with it. For example, one may want to model that a program is running on an operating system without modeling the complete operating system or all the program's interactions with it. Likewise, one may want to model that a system is interacting with a user, for example, to understand what inputs are provided by the user.

Solution: The key approach to this is the introduction of an environment. Depending on the type of environment, this may be represented as a prov:Agent, prov:Activity, or prov:Entity. There may be connections between the system's components and the environment.

Scenario: Take the example of nowpeople:Alice from the data journalism scenario. She acts as the environment for all the various tools that she used to create the figure included in the article. To show this, we link each activity (expressing the execution of a tool) to her.

```
is:download1 prov:wasAssociatedWith nowpeople:Alice.
is:unzip2 prov:wasAssociatedWith nowpeople:Alice.
is:xls2csv3 prov:wasAssociatedWith nowpeople:Alice.
is:triplify4 prov:wasAssociatedWith nowpeople:Alice.
is:extractor5 prov:wasAssociatedWith nowpeople:Alice.
```

Discussion: Like using a general entity to represent a file in the revision recipe (Recipe 4.1.4), the introduction of an environment provides bounds within which provenance can be described. The definition of such bounds is an important modeling principle in provenance systems as it gives a scope to provenance. Oftentimes, these groupings of provenance can be encapsulated within bundles. For example, all the provenance associated with Alice or all the provenance produced by an organization can be grouped together.

4.1.9   MODELING SUB-ACTIVITIES

Question: How does one model activities composed of other activities?

Context: Many times a long running activity is composed of separate individual activities. There is no specific construct within PROV for modeling the relationship between activities.

Solution: Here are four steps for modeling sub-activities.

1. Introduce the individual sub-activities. Ensure that they are within the time boundaries of the larger activity.

2. Use dct:hasPart to link the sub-activities to the larger activity. The property dct:hasPart belongs to the widely used Dublin Core Terms vocabulary and defines that a resource is either physically or logically included in something else.

3. Model explicitly the exchange of entities between the sub-activities.

4. Optionally, use specialization to model when the output or input of the larger activity is the output or input to the sub-activities.

   For example, we can model that downloading, unzipping, and converting an Excel file are all sub-activities of file preparation as follows:

```
:filePrep a prov:Activity .
:filePrep prov:used :fileToPrep.
:preparedFile prov:wasGeneratedBy :filePrep.

:filePrep dct:hasPart :download1, :unzip2, :xls2csv3 .

#connect the sub-activities together
:download1 prov:used :zipFile .
:f1 prov:wasGeneratedBy :download1.
:unzip2 prov:used :f1 .
:f2 prov:wasGeneratedBy :unzip2 .
:xls2csv3 prov:used :f2 .
:f4 prov:wasGeneratedBy :xls2csv3 .

:zipFile prov:specializationOf :fileToPrep .
:f4 prov:specializationOf :preparedFile .
```

Scenario: Again this could be applied to group the activities of Alice's preparation of the employment graph together. This could also be used to group all the activities involved in the creation of the employment article. These groupings facilitate the retrieval of provenance associated with a data item. For example, an auditor could focus on the provenance of a specific article without writing a sophisticated query.

Discussion: For complex activity structures, it may be useful to use the prov:hadPlan relation to link to a more complex workflow or plan based representation. These plans can give additional insights into the relationships between activities.

4.2 ORGANIZING

One challenge is organizing provenance in such a way that it is easy to query, find, and manage. In this section, we describe several recipes for organizing provenance.

4.2.1 STITCH PROVENANCE TOGETHER

Question: How does a consumer of provenance trace back information across multiple components, computers, or organizations?

Context: The provenance of any given data item can extend across multiple programs, computers, and even organizations. How does one ensure that the provenance can be readily traced and queried?

Solution: There are two key rules to enable provenance to be stitched together:

1. **Always assert relationships**. In provenance, there should be no entity, activity, or agent that stands on its own. They should always be connected to something else by a relation even if that relation points to a component on another system.

2. **Adopt the same URIs for the same entities across components**. To allow provenance to be connected across systems, components need to be able to refer to things produced by other components. There are two approaches to ensure that components have common identifiers:

   - Use a shared convention for URIs at design time. By defining a convention, components use the same identifier space to refer to entities. For example, in this book, we use http://www.provbook.org/is/# as a convention to refer to provenance specific identifiers.
   - Send URIs between components. Components can transmit identifiers for entities when communicating. This can be done in the headers of messages or even embedded in the content itself [17]. When designing a system, it is good to agree on where to locate these shared identifiers.

Scenario: If NowNews adopts such a convention and propagates its URIs with its content, the provenance of the policy report produced by PolicyOrg can be traced back to the provenance of the NowNews employment article, which the report depends on.

Discussion: By adopting common identifiers, one can help ensure that the provenance graph being constructed stays connected across different components. Note, we recommend using URLs that are dereferenceable.

4.2.2 USE CONTENT-NEGOTIATON WHEN EXPOSING PROVENANCE

Question: What serialization should I provide provenance in?

Context: Depending on the application, it may be beneficial to have the provenance of a data item in a certain format. For example, a Web application may need provenance in JSON, a data integration application may need it in RDF, while for a quick look at provenance, a graphical representation as an SVG may be more appropriate.

Solution: PROV was designed to supply provenance in multiple representations catering for different types of development platforms whether they are enterprise XML-based applications or RDF-based Semantic Web applications. Thus, data providers should ensure their provenance is useful for a variety of needs by supplying the same provenance data in multiple representations using content-negotiation.[2] For example, the provenance of this book is located at `http://www.provbook.org/provenance`. The following lists the curl commands to retrieve 4 different representations of the same provenance (namely, TURTLE, SVG, XML, and JSON).

```
curl -L http://www.provbook.org/provenance -sH "Accept: text/turtle"
curl -L http://www.provbook.org/provenance -sH "Accept: image/svg+xml"
curl -L http://www.provbook.org/provenance -sH "Accept: application/provenance+xml"
curl -L http://www.provbook.org/provenance -sH "Accept: application/json"
```

Figure 6.12 shows a choice of 8 different representations being supported by the `http://www.provbook.org` website.

The provenance information corresponding to a particular format can either be generated up-front or produced on the fly. Most web servers and environments support content negotiation. Note that adopting content negotiation influences the design of URLs, as follows. We suggest using a URL without extension to denote the resource[3] for the provenance data, irrespective of its serialization. Dereferencing this URL redirects to a URL with an extension for a concrete representation. See Section 6.3.4 for an example from the provenance of this book.

Scenario: By offering multiple provenance representations, NowNews can support different developers in using their content. For instance, Alice may be more comfortable in using the RDF representation whereas Nick may find JSON more useful.

Discussion: This recipe does have some set-up cost in terms of generating more types of provenance and configuring a web server appropriately. However, for public services, this extra cost may be beneficial in terms of providing easy access. Tools exists (see Chapter 6) to generate such representations automatically.

4.2.3   BUNDLE UP AND PROVIDE ATTRIBUTION TO PROVENANCE

Question: How can a consumer of provenance determine the asserter of provenance?

Context: There is usually not an authoritative source of provenance, so provenance can be discovered in multiple places. Thus, information about the asserter of provenance, and more generally, provenance of provenance become crucial to assess the quality of provenance.

---

[2]`http://www.w3.org/QA/2006/02/content_negotiation.html`
[3]Such a resource is usually termed non-information resource.

Solution: PROV provides a mechanism, referred to as bundle (see Section 3.7), by which a set of provenance assertions can be packaged up and named. A bundle is also an entity, so its provenance can be expressed using PROV mechanisms and by referring to its name.

In the following RDF snippet, we use the TRIG notation to represent a bundle as a named graph called provapi:d001,[4] which contains provenance assertions describing what Alice did.

```
provapi:d001 {
          <ftp://ftp.bls.gov/pub/special.requests/oes/oesm11st.zip> a prov:Entity , void:Dataset ;
               rdfs:label "employment-stats-2011" .

     is:f1 a prov:Entity ;
               rdfs:label "local copy of zip" .

     is:f1 prov:wasDerivedFrom <ftp://ftp.bls.gov/pub/special.requests/oes/oesm11st.zip> .

     # ... more triples
}
```

In this book, we adopt the convention that the identifier of a bundle, such as provapi:d001, can be dereferenced to obtain a representation of the bundle. Content negotiation is also applicable here to retrieve the bundle in the desirable representation. For instance, the above named graph is returned when the application/trig mime type is requested.[5]

Having created a bundle, the provenance of provapi:d001 can then be expressed, here represented in a separate document, stating that provapi:d001 is an entity bundle attributed to Alice.

```
{
     provapi:d001 a prov:Entity , prov:Bundle ;
               prov:wasAttributedTo nowpeople:Alice .
}
```

Scenario: In the above example, we can see that the provenance is attributed to particular people or agents within the scenario.

Discussion: It is good practice to bundle up provenance assertions, and provide their provenance. Minimally, provenance of provenance should include attribution, but the full PROV model is available to express it. The next question is: where do we find the provenance of a bundle such as provapi:d001? For answers, see "Embedding Provenance" (Section 4.2.4), "Provenance in HTTP Headers" (Section 4.2.6), and "Self-Attribution" (Section 4.2.7).

4.2.4   EMBEDDING PROVENANCE IN HTML

Question: How do I make provenance available in a web page?

---

[4]We use the abbreviation provapi:d001 to denote the URI http://www.provbook.org/provapi/documents/d001.
[5]If the requested mime type does not allow the encoding of the bundle construct itself (for instance, TURTLE has no support for named graph), only its content is returned. Thus, such a kind of representation is degraded, since it lost information. Hence, we record this with a lower quality in the content negotiation (qs=0.5). (See Figure 6.12 for an illustration on www.provbook.org.)

Context: PROV is designed to interchange provenance on the Web. Given a web page, there are multiple options for exposing its provenance and there are some issues to be conscious of.

Solution: Before applying any of the options below, we suggest that all parts of the web page for which provenance needs to be given have well-defined identifiers. This can be achieved by placing the relevant portion of the page in a `div` tag and giving it a specific id tag. This means that it can be referred to specifically. There are two major options for embedding provenance. One is "by value" and the other is "by reference." With "by value," we mean that a page can be marked-up using RDFa with PROV statements directly. This is discussed in more detail in Chapter 6. As far as "by reference" is concerned, PROV-AQ [25] defines mechanisms for referring to provenance either stored as a separate file on the Web or stored in a specific service for maintaining provenance often called a provenance store.

To refer to a resource containing provenance, we add a `prov:has_provenance` and `prov:has_anchor` statements to the head of the HTML file. For example, we can link to the provenance of the `www.provbook.org` homepage.

```
<head>
    <link rel="http://www.w3.org/ns/prov#has_provenance"
          href="http://www.provbook.org/provenance">
    <link rel="http://www.w3.org/ns/prov#has_anchor"
          href="http://www.provbook.org/">
</head>
```

The object of a `prov:has_provenance` property points to the resource containing the provenance, whereas that of `prov:has_anchor` points to where to locate the actual entity being described within that provenance. We note that without using the relation `prov:has_anchor`, the anchor defaults to using the URL of the current page. However, in many instances, the URL for the entity within the provenance may be different than the page, for example, when referring to a time and date stamped version of the page. The same pattern applies for referring to provenance stored in a provenance store, except instead of using `prov:has_provenance`, one uses `prov:has_query_service` to denote the location of store.

Scenario: By exposing the provenance of its data, NowNews enables PolicyOrg to use content that it perceives as quality (i.e. based on the fact that it is based on a government source).

Discussion: There are obvious trade-offs between embedding the provenance directly in the HTML using RDFa or referring to it externally. One approach is to store most provenance in a separate provenance store and then expose simple statements (e.g., a paragraph was quoted from a website) using RDFa to enable discovery of its provenance.

4.2.5 EMBEDDING PROVENANCE IN OTHER MEDIA

Question: Right, I know how to embed provenance metadata in html, but what about other formats, such as JPG, PDF, or SVG?

Context: The W3C Provenance Working Group did not suggest a specific way to embed provenance metadata in non-HTML formats. However, it sets out a philosophy that can be adapted for
each technology.

Solution: The solution consists in adding metadata fields prov:has_anchor,
prov:has_provenance, and prov:has_query_service to a document, reusing metadata
capabilities supported by the specific format.

In this book, we leverage the Extensible Metadata Platform[6] (XMP) to support provenance
metadata. XMP can be used with a variety of file formats, including JPG and PDF. Such metadata
can be created and retrieved using the open source software ExifTool.[7] For example, provenance
metadata is serialized in RDF/XML format, as follows.

```
exiftool -b -XMP plot.jpg

<?xpacket begin=' ' id='W5M0MpCehiHzreSzNTczkc9d'?>
<x:xmpmeta xmlns:x='adobe:ns:meta/' x:xmptk='Image::ExifTool 9.27'>
 <rdf:RDF xmlns:rdf='http://www.w3.org/1999/02/22-rdf-syntax-ns#'>

  <rdf:Description rdf:about=''
     xmlns:prov='http://www.w3.org/ns/prov#'>
      <prov:has_anchor>http://www.provbook.org/nownews/plot.jpg</prov:has_anchor>
      <prov:has_provenance>http://www.provbook.org/provenance</prov:has_provenance>
  </rdf:Description>
 </rdf:RDF>
</x:xmpmeta>
```

The Scalable Vector Graphics SVG format defines a metadata element, in which metadata
can also be serialized in RDF/XML format. An SVG file with provenance metadata looks as follows.

```
<svg>
  ...
   <metadata>
     <rdf:RDF xmlns:rdf = "http://www.w3.org/1999/02/22-rdf-syntax-ns#"
            xmlns:prov = "http://www.w3.org/ns/prov#" >
      <rdf:Description about="">
         <prov:has_anchor>http://www.provbook.org/provapi/documents/d000.svg</prov:has_anchor>
         <prov:has_provenance>http://www.provbook.org/provenance</prov:has_provenance>
      </rdf:Description>
     </rdf:RDF>
   </metadata>
  ...
</svg>
```

Scenario: Using these techniques, Alice can embed the provenance of the figure she created
allowing its provenance to be retrieved, for example, in the licensing use case.

Discussion: For full interoperability, a standardization activity is required to define the various
metadata fields for the various file formats. The proposal outlined above is a reasonable input

---

[6]XMP: http://en.wikipedia.org/wiki/Extensible_Metadata_Platform
[7]ExifTool: http://www.sno.phy.queensu.ca/~phil/exiftool/

to such standardization activity. Its main benefit is its strong alignment with the HTML way of embedding provenance metadata.

4.2.6 WHEN ALL ELSE FAILS, ADD PROVENANCE TO HTTP HEADERS

Question: How can we communicate provenance if it is not suitable to embed it in a resource representation?

Context: In some cases, it is not suitable to embed provenance in a resource representation (as suggested by Recipe 4.2.4), because either we do not have control over the contents of the resource representation or the representation does not allow for metadata to be stored.

Solution: PROV allows for provenance-related information to be embedded in HTTP headers. In this case, provenance cannot be passed by value; instead, it is passed by reference, by means of a URI pointing to a provenance resource, or to a provenance service.

The following example[8] shows a Link header field, which links provapi:d001 to provapi:d002, with the relation prov:has_provenance.

```
curl -s -D - http://www.provbook.org/provapi/documents/d001 -o /tmp/foo -H "Accept: application/trig"

HTTP/1.1 200 OK
Date: Wed, 26 Jun 2013 21:57:56 GMT
Server: Apache/2.2.3 (Red Hat)
Content-Location: d001.trig
Vary: negotiate,accept
TCN: choice
Last-Modified: Wed, 26 Jun 2013 21:41:04 GMT
ETag: "18b808f-1350-4e01580a3dc00;18b8091-187-4e01591915b00"
Accept-Ranges: bytes
Content-Length: 4944
 Link: <http://www.provbook.org/provapi/documents/d002>;
        rel="http://www.w3.org/ns/prov#has_provenance";
        anchor="http://www.provbook.org/provapi/documents/d001.trig"
Connection: close
Content-Type: application/trig
```

The bundle provapi:d002 is now represented as follows:

```
provapi:d002 {
        nowpeople:Alice a prov:Agent .

        provapi:d001 a prov:Entity , prov:Bundle ;
                prov:wasAttributedTo nowpeople:Alice .
}
```

Scenario: This approach could be useful for the GovStat to expose the provenance of its statistics made available as zip files.

---

[8]For the purpose of presentation, the headers shown in this Chapter are simplified and display a single Link header field; we refer the reader to Section 6.3 for an explanation about the multiplicity of this field.

Discussion: This approach is particularly relevant for content providers that do not have the possibility of embedding provenance directly in the resource representations they serve. However, this approach can present technical challenges for HTTP clients. For instance, JavaScript embedded in an HTML page is typically not given access to the HTTP header of the response that contained the document currently displayed. Plugins and browser-specific solutions may need to be developed, which hampers portability.

Furthermore, given a bundle identifier, one can also dereference the identifier with an HTTP GET request (or alternatively we can use an HTTP HEAD request), and obtain the location of its provenance in the Link field. Inevitably, the same question arises: what is the provenance of this latter provenance? Recipe "Self-Referential Bundle" (Section 4.2.7) helps address this potentially infinite situation.

4.2.7   EMBEDDING PROVENANCE IN BUNDLES: SELF-REFERENTIAL BUNDLES

Question: How do we avoid an infinite chain of bundles, each expressing the provenance of a previous bundle in the chain?

Context: The provenance of a bundle is given by some provenance statements; these statements can be packaged up and named, resulting in a new bundle, whose provenance can also be expressed. This situation is reminiscent of certificate authorities, which attest the validity of certificates published by other authorities. The highest certificate authorities use self-signed certificates (i.e. certificates signed with their own private keys).

Solution: By defining a self-referential bundle, which includes its own provenance, one can avoid infinite provenance chains.

The following bundle `provapi:d003` describes the provenance of `provapi:d002`, but it is also self-referential, since it contains its own provenance.

```
provapi:d003 {
        nowpeople:Alice a prov:Agent .

        provapi:d002 a prov:Entity , prov:Bundle ;
               prov:wasAttributedTo nowpeople:Alice .

        provapi:d003 a prov:Entity , prov:Bundle ;
               prov:wasAttributedTo nowpeople:Alice .
}
```

The Link header field when dereferencing `provapi:d002` is as follows:

```
Link: <http://www.provbook.org/provapi/documents/d003>;
      rel="http://www.w3.org/ns/prov#has_provenance";
      anchor="http://www.provbook.org/provapi/documents/d002.trig"
```

while the one for `provapi:d003` refers to itself:

```
Link: <http://www.provbook.org/provapi/documents/d003>;
      rel="http://www.w3.org/ns/prov#has_provenance";
      anchor="http://www.provbook.org/provapi/documents/d003.trig"
```

Scenario: This can be used for members of NowNews to attest to the set of provenance created by their actions. For example, Alice creating the figure and the attributing the provenance to herself.

Discussion: Self-referential bundles are self-contained, and may appear attractive to HTTP clients, since clients do not require any further GET request to download further provenance. In practice, however, nothing guarantees that the self-contained provenance is the only provenance for a given bundle: other providers, or even the same provider, may provide further provenance in other bundles. Clients should assume this is the case, and thus should retrieve provenance from all possible sources. There is not one recommended technique to find all possible sources for provenance. In a SPARQL context, SPARQL queries aiming to process provenance of bundles should be designed so that bundle entities are looked up across multiple named graphs. Alternatively, we could imagine that search engines, or dedicated "provenance registries," could be queried to find further provenance.

4.2.8   WHEN DISPLAYING PROVENANCE, ADOPT CONVENTIONAL LAYOUT

Question: How does one display provenance graphs?

Context: Frequently, it is helpful to visualize provenance graphs, in particular, for pedagogical, debugging, or communication purposes (e.g., in slides).

Solution: To ensure that people who are familiar with PROV can quickly grasp a visualized provenance graph, the W3C Provenance Working Group has developed a style guide for drawing PROV graphs, which is available at `http://www.w3.org/2011/prov/wiki/Diagrams`. This style has been adopted throughout this book.

Scenario: In the scenario, this could be used by Nick, the webmaster, in the licensing use case to view the provenance of various data sets.

Discussion: It is important to realize that this visual style is meant to make it easy to understand provenance graphs from the data structure point of view. Java developers, for instance, expect to read code adopting a common layout convention. Likewise, PROV developers expect to see provenance using this convention. In many cases, in particular when displaying results to users, it may be best to develop application-specific visualizations that can abstract away from the complexity of the PROV data structure and instead display what is appropriate for the user. For example, in an application focused on timeliness, one may just list containing activities, and their start and end times.

4.3 COLLECTING

4.3.1 USE STRUCTURED LOGS TO COLLECT PROVENANCE

Question: What is a simple way to add provenance to a system?

Context: In many cases, one wants to quickly add provenance to a program or other system. Normally, we suggest instrumenting the system with specialized provenance middleware to capture and record provenance. However, in some cases this may not be possible.

Solution: Most programming environments support logging through libraries like log4j (See http://logging.apache.org). By systematically generating log files that incorporate dependencies between entities, the start and stop time of activities, and the inputs and outputs of activities, one can write a program to extract provenance from these log files. An example of generating structured logs can be found at http://netlogger.lbl.gov/ [37].

Scenario: This approach could be used within NowNews to implement their provenance collection strategy.

Discussion: Building on existing logging infrastructure is a straightforward way of adding provenance. In general, there are many existing pieces of infrastructure that can be repurposed for collecting provenance. For example, in prior work, we used web service adaptors to capture provenance [16].

4.3.2 COLLECT IN A LOCAL FORM, EXPOSE AS PROV

Question: In what format should one collect provenance information?

Context: Systems may be built on different technologies. Data integration applications may use RDF, web applications may use SQL, and enterprise applications may use parts of the Java stack. The question then arises as to what infrastructure should be used to collect and store provenance.

Solution: We suggest the following steps for designing a collection procedure for an application.

1. Design the collection and storage system for provenance in a way that fits with application requirements. For example, it may be useful to store provenance in a SQL database, or represent it using logging statements.

2. Ensure that provenance can be *converted* to PROV extending PROV concepts where needed for the application.

3. Convert provenance from its internal representation to PROV, and expose it as discussed in Section 4.2.2.

Scenario: NowNews could extend its existing systems for storing articles to store provenance information.

Discussion: The key focus of this recipe is on PROV as an interoperability mechanism. Applications should not be afraid to capture and collect provenance in the manner that is best for their applications but then to provide access to that data set using these recommendations.

4.4   ANTI-PATTERNS

This section catalogs some problems and shortcomings we commonly see in provenance graphs.

4.4.1   ACTIVITY BUT NO DERIVATION

Question: Is there a connection between an entity generated by an activity and the entities used by that activity?

Context: Entities generated by an activity may not be dependent on all of the entities used by that activity. Figure 4.1 illustrates the issue. Articles 1 and 2 are inputs to the same editing activity, and it is unclear whether the resulting edited articles are dependent some or all the input articles. Unfortunately, provenance graphs found in the wild are generated under the unsafe assumption that all entities generated by an activity are dependent on all input entities.

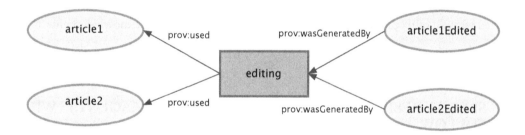

Figure 4.1: An editing activity over two articles. Is `article1Edited` dependent on `article1`, `article2`, or both?

Solution: There are two solutions to this problem. One is to introduce explicit derivation relations between entities that are dependent on one another. In this case, it would be between the edited article and the corresponding input article. The other is to define an application-specific extension to `prov:Activity` (e.g., Function) that defines each output as being derived from all inputs. The first solution is somewhat preferable as it makes the derivations explicit and is more interoperable.

Scenario: As shown in the example above, following this recipe for example ensures that edited articles are correctly associated with their input articles.

Discussion: This problem can be avoided by following the process of going from data flow to process flow.

4.4.2   ASSOCIATION BUT NO ATTRIBUTION

Question: Are all the entities generated by an activity attributed to the agents that are associated with that activity?

Context: Entities generated by an activity are not automatically attributed to the agent who is responsible for that activity. For example, an article publication activity may be the responsibility (i.e. associated with) of an editor but the article produced may be attributed to the author of that article.

Solution: Similar to Recipe 4.4.1, a solution is to explicitly express the attribution from the entities to the agent that is responsible for them. In some cases, it may be useful to use prov:actedOnBehalf to express chains of responsibility to reduce the number of prov:wasAttributedTo edges linking to one entity. For example, if an article is attributed to the author who acted on behalf on an editor, then the editor shares some responsibility for the article as well.

Scenario: By applying this recipe, NowNews can ensure that credit (and blame) is correctly associated with all of its products, helping it to address Use Case 2.7 Acknowledgments.

Discussion: Note that by following Recipe 4.1.3, this problem is prevented as attribution is introduced before association.

4.4.3   SPECIFY RESPONSIBILITY FIRST, WHAT A PROV:AGENT IS WILL FOLLOW

Question: Is this person, organization, or software really a prov:Agent?

Context: Other definitions of agent (e.g., prov:foaf:Agent) differ from prov:Agent. This sometimes leads to people, organizations, and software being defined as prov:Agent when they are not meant to be. In particular, a person may not have responsibility, in the PROV sense, for something. Being a prov:Agent implies responsibility for some entity, activity, or agent.

Solution: To avoid confusion, we suggest only applying the type prov:Agent after associating a thing to an activity or entity using a responsibility property such as prov:wasAttributedTo or prov:wasAssociatedWith, respectively. The property prov:actedOnBehalfOf can also be used to create responsibility chains between agents. In this fashion, one can be sure that the semantics are appropriately assigned.

Scenario: By applying this recipe, NowNews can correctly ensure that responsibility is given to the correct parties, namely the editors and authors. This would help address Use Case 2.9

Publication Embargo, which requires attribution to find who was responsible for the release of data.

Discussion: PROV concepts are designed to be easily and quickly applicable but it is also useful to be aware of their semantics and use them appropriately.

4.5 SUMMARY

This chapter provided a collection of recipes that can be applied to create better PROV or to clarify how to use PROV in different situations. We encourage readers to let us know if they have additional recipes that would be of use.

CHAPTER 5

Validation, Compliance, Quality, Replay

Chapter 3 defines the PROV ontology, but sets no constraint on its use. Hence, it is possible to express provenance descriptions that describe events that could not possibly happen in the world as we know it. How should we interpret a description in which the end of an activity precedes its start? What about a description in which an entity is used before it was created? Such descriptions are usually not desirable, since they provide an inconsistent history of events. It may, therefore, come as a surprise to the reader that the PROV-O ontology and the PROV data model[1] allow for such descriptions to be expressed. This design choice is intentional to ensure that the use of PROV remains as easy as possible.

To address this concern, the PROV ontology is accompanied by a set of constraints [8] that *valid provenance* is expected to satisfy. Using this set of constraints, validators for provenance have been implemented[2] and deployed.[3] They allow users to determine whether provenance descriptions are valid. Provenance descriptions are said to be valid if they present a consistent history of objects and their interactions.

Having determined whether a provenance graph is valid is a first, domain-independent analysis, which is a preliminary to its deeper utilization. Chapter 2 introduced various kinds of provenance utilization: compliance, assessing the quality of artifacts, cataloging, and replaying past executions. We explore technical requirements underpinning such functionality, in the data journalism scenario, and explain how this functionality can be implemented using simple queries over provenance.

This chapter is organized as follows. First, we introduce a few validation use cases (Section 5.1), which we follow with the key principles of validation (Section 5.2). Requirements for provenance utilization are then exposed in Section 5.3, and followed by implementations using Semantic Web technologies in Section 5.4. Readers interested in these specific examples can skip Section 5.2, which provides the more detailed underpinnings of generic provenance validation.

---

[1]It is an open question as to whether validity constraints could be expressed in OWL2. The provenance working group decided not to axiomatize these constraints in the ontology, to ensure that PROV-O conforms to the OWL-RL profile to allow efficient implementations.

[2]prov-check: `https://github.com/pgroth/prov-check`

[3]ProvValidator: `https://provenance.ecs.soton.ac.uk/validator`

5.1 VALIDATION USE CASES

In this section, we present a few situations in which there exist inconsistencies in provenance descriptions, which a provenance validator can identify. Our assumption throughout this chapter is that intent and behavior of the provenance provider are not malicious, and that provenance has been recorded faithfully; however, provenance descriptions may not have been expressed suitably, which can result in issues like those we expose here.

Use Case 5.1 (Version Issue)   NowNews provenance indicates that an article includes a quote from another document. That document happens to be the compilation produced by PolicyOrg. The compilation itself includes the NowNews article. A circularity occurs in the provenance, which is indicative of a problematic description.

The problem of circularity in provenance may be due to different reasons. It may be symptomatic of a problem with the descriptions of different versions of a document. For instance, let us assume that two versions of the compilation exist. A NowNews article includes a quote taken from the first version, but the NowNews article is in turn included in the second version of the compilation. The provenance description becomes problematic if it refers to the compilation in general, irrespective of its actual version, as illustrated in Use Case 5.1.

Use Case 5.2 (Date Issue)   A plot was computed from a data set. The plot's timestamp is found to precede the data set's timestamp. Given that the plot was derived from the data set, this is an indicator of a problem.

Something cannot be used before it is generated: a data set cannot be used to produce a plot before the data set is generated. Assuming that the graph topology is correct, i.e. that the plot is indeed derived from the data set, then time annotations may be the source of the problem. An explanation for the problem may be that Alice had just installed a new computer, but had not configured the computer's internal clock properly, and as a result the plot was annotated with an incorrect generation time.

5.2 PRINCIPLES OF VALIDATION

The intent of validation is to ensure that PROV descriptions represent a consistent history of objects and their interactions. Once established, valid PROV descriptions are safe to use for the purpose of logical reasoning and other kinds of analysis [8], such as the ones described in the rest of this chapter.

Time is critical for provenance, since it can help corroborate provenance descriptions. However, PROV makes no assumption on the clocks being used when asserting time: a unique clock is not expected, nor clocks are expected to be synchronized. Instead, PROV relies on an event model describing changes in the world. PROV defines five types of events: generation, invalidation, and usage of entities, and start and end of activities. Furthermore, two partial orders between events have been defined: "precedes" and "strictly precedes." Ordering inequalities of the form "*event1 (strictly) precedes event2*" can be inferred from PROV descriptions according to inferences defined in [8]. Provenance descriptions are valid if they satisfy event ordering inequalities that can be inferred. The event ordering constraints, as well as a few other constraints in [8], are application-independent.

Given this event model, time information is seen as an annotation of events: each event occurs at a specific point in time. With application-specific knowledge about clocks, applications can also use such time information to check whether time information ordering is compatible with event ordering.

It should be noted that most applications are not expected to implement PROV-CONSTRAINTS. Instead, these constraints are to be implemented by validators. To understand the results produced by validators, and correct invalid provenance, it is useful for developers to understand the validation routine, which we now explain. The process of validation, i.e. checking that provenance is valid, consists of three steps.

1. normalize descriptions;
2. generate ordering inequalities;
3. check for absence of undesirable cycles.

Normalization is the process by which some basic inferences are performed, uniqueness is checked, and existential identifiers are introduced where appropriate. Normalization, whose technical details can be found in [8, section 7], aims to provide a systematic representation of provenance descriptions conducive to reasoning. Ordering inequalities can then be inferred by a set of inference rules listed in [8, section 6]. A cycle of ordering inequalities is undesirable when it contains a "strictly precedes" relation.

In the rest of this section, we provide some intuition about events and ordering inequalities that are implied by PROV descriptions. Equipped with this knowledge, we then revisit the use cases of Section 5.1.

5.2.1   EVENTS AND THEIR ORDERING

In PROV, entities and activities have a lifetime, delimited by two events: an entity's lifetime begins with a generation event and terminates with an invalidation event. Likewise, an activity's lifetime begins with a start event and terminates with an end event. A further event marks the usage of an entity by an activity.

To understand events and their ordering, we use *provenance event charts*, as illustrated in Figure 5.1. In a provenance event chart, the horizontal axis represents an event line, i.e. a line in which older events appear to the left and more recent events appear to the right. Each event is marked by a dotted vertical line.

In Figure 5.1, an entity :e has a lifetime delimited by its generation :g and its invalidation :i. As in Chapter 3, we use the green inverted triangle to represent an event, and link it to the associated entities and activities: so the generation event :g links entity :e and activity :a1; likewise, the invalidation event :i links :e and :a2. A first constraint is that generation is expected to precede invalidation for entity :e. A second constraint is that, for every usage :u of entity :e, this usage event is expected to occur after the generation of the entity, and before its invalidation. The constraints implied by :g, :u, and :i are summarized in the box to the right of Figure 5.1. (In Figure 5.1, activities :a1, :a2, and :a3 need not be distinct.)

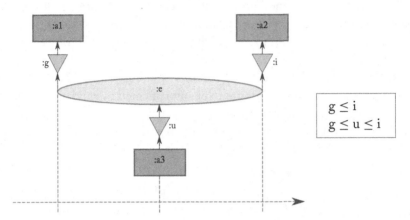

Figure 5.1: Provenance Event Chart for Entity: generation of an entity precedes its invalidation. Generation of an entity precedes any usage of the entity, which precedes its invalidation.

A provenance event chart for activity :a is illustrated in Figure 5.2. There, we see that the start event :s is expected to precede the end event :e.

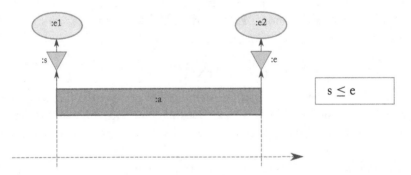

Figure 5.2: Provenance Event Chart for Activity: start of an activity precedes its end.

For an activity to be involved in a generation or usage event, this activity must exist: hence, usage and generation should occur during the lifetime of the activity. This ordering constraint is

illustrated by the provenance event chart in Figure 5.3, which displays a usage :u3 and a generation :g4, which follow the start :s and precede the end :e of the activity.

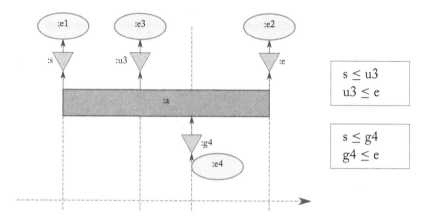

Figure 5.3: Provenance Event Chart for Usage and Generation: a usage or generation of an entity by
an activity occurs during the lifetime of the activity.

The previous ordering constraints were of the form $evt_1 \leq evt_2$, which allows both events
to occur simultaneously (See Section 5.2.2). Derivation is the only PROV construct from which
we can infer a strict ordering, as illustrated in Figure 5.4. There, we see a fully qualified derivation
:d from entity :e2 to entity :e1. The generation :g1 of :e1 is required to strictly precede the
generation :g2 of :e2, since the entity :e2 is expected to be newer than :e1.

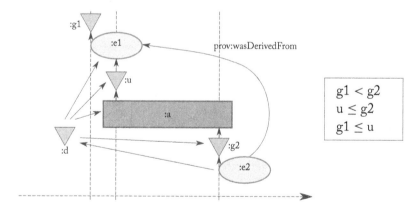

Figure 5.4: Provenance Event Chart for Derivation: if an entity is derived from another, the generation of the latter strictly precedes the generation of the former.

5.2.2   SIMULTANEOUS EVENTS

Coming back to the data journalism example of Chapter 2, the article on employment data was generated by the `writeArticle` activity, under the control of Bob. However, the `writeArticle` activity consists of several subactivities, one of which, the publication activity pub, performed by Nick the webmaster, also generated the article. So, the article is effectively generated by two activities `writeArticle` and pub. This is a common pattern in provenance traces, in which a subactivity and its parent activity both generate an entity. Thus, it would not make sense if those two generation events occurred at different instants.

This situation is illustrated by the Provenance Event Chart in Figure 5.5, displaying an entity :e generated by two different events :g1 and :g2. Both events are expected to occur simultaneously, which is encoded by the pair of constraints $g1 \leq g2$ and $g2 \leq g1$.

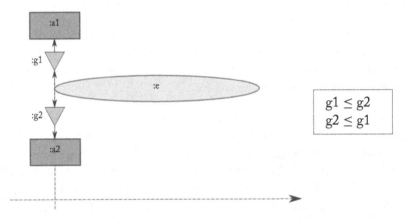

Figure 5.5: Provenance Event Chart for Simultaneous Generations: two distinct generations of the same entity by two activities that occur simultaneously.

Similar constraints of simultaneity apply whenever we are in the presence of multiple invalidations of the same entity, multiple starts of the same activity, and multiple ends of the same activity.

Remark 5.3   While an entity may be generated (resp. invalidated) by multiple activities, with the simultaneity constraint described in Section 5.2.2, it can be generated (resp. invalidated) by one given activity at most once. Likewise, while an activity may be started (resp. ended) by multiple starter activities (resp. ender activities), it can be started (resp. ended) by one given starter activity (resp. ender activity) at most once. Such constraints are referred to as uniqueness constraints [8, section 6.1].

5.2.3   NESTED INTERVALS AND SPECIALIZATION

If an entity :e1 specializes an entity :e, then :e1 "inherits" all the attributes of :e, and also presents extra specific attributes. Given this, the lifetime interval of :e1 is necessarily nested in the lifetime interval of :e. Indeed, the attributes of :e are known to be fixed for the lifetime of :e; furthermore, the extra specific attributes of :e1 may only hold for a shorter interval. If two entities :e1 and :e2 are alternates of each other, there is no constraint on their lifetime intervals; furthermore, they may or may not overlap.

These constraints are summarized by the Provenance Event Chart of Figure 5.6, which presents an entity :e, and two specializations :e1 and :e2, which consequently are alternates of each other.

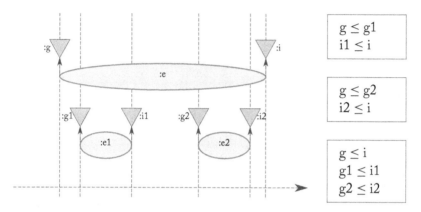

Figure 5.6: Provenance Event Chart for Specialization: the lifetime intervals of specializations :e1 and :e2 are included in the lifetime interval of the more general entity :e; in addition, :e1 and :e2 are alternate and, thus, unconstrained.

5.2.4   USE CASES REVISITED

We now revisit Use Case 5.1 (Version Issue), and illustrate it with the Provenance Event Chart in Figure 5.7. First let us consider a valid description of this use case: we have (1) an article :art that includes a quote from a compilation :c1, and (2) a compilation :c2 that is a revision of :c1, including article :art. Compilations :c1 and :c2 are specializations of :c, the compilation of articles irrespective of the actual version.

By successive application of constraints for derivation (see Figure 5.4), we obtain that generation of :c1 strictly precedes that of :art, and generation of :art strictly precedes that of :c2. Furthermore, given that :c1 and :c2's lifetime is nested inside that of :c, the generation of :c precedes that of :c1 and :c2.

Use Case 5.1 (Version Issue) does not refer to different versions of the compilation, but a single one :c. Article :art, which includes a quote from :c, is itself included in :c. In this case,

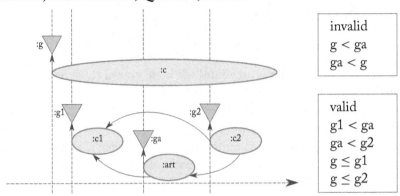

Figure 5.7: Event Chart for Use Case 5.1: article :art includes quotes from :c1 and was itself included in :c2. The generation :g1 of :c1 strictly precedes the generation :ga of :art, which strictly precedes the generation :g2 of :c2

we obtain that generation of :c strictly precedes that of :art, and generation of :art strictly precedes that of :c, which is impossible.

Remark 5.4   Provenance validation helps identify problems in provenance descriptions, but it is up to the asserter to rectify invalid provenance descriptions. For instance, another explanation for Use Case 5.1 is that article :art includes a quote from article :art2, and both articles are included in compilation :c2.

We now revisit Use Case 5.2 (Date Issue), and again depict valid provenance for it. A plot :plot was derived from a data set :data. By application of the constraints associated with derivation, we infer that the generation of the data set strictly precedes the generation of the plot. In this use case, events are annoted with time information, written along the event line, for each event, in Figure 5.8. If, in this application, we make the assumption that clocks are properly synchronized, then the *time* at which the data set is generated should also strictly precede the *time* at which the plot was created.

In Use Case 5.2 (Date Issue), time information is found not to satisfy this ordering constraint. An explanation for this situation is that the synchronized clock assumption does not hold here, since Alice has just installed a computer without configuring the clock properly.

5.3   UTILIZING PROVENANCE

We explore the four broad categories of provenance usage introduced in Chapter 2: *quality assessment*, *compliance*, *cataloging*, and *replay*. From the use cases, we extract technical requirements that

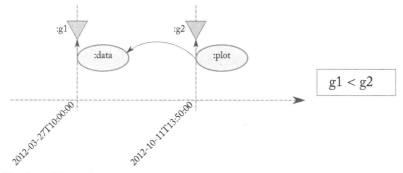

Figure 5.8: Checking Time Annotation.

make explicit some functionality a tool could support. We then explain how these requirements can be addressed using provenance in Section 5.4.

In the context of this book, we define *compliance check* as the process of determining whether some past execution, described by some provenance, is compatible with some constraints, rules, or policies. We assume a faithful and accurate description of past executions. Hence, by checking provenance descriptions, we can determine whether past executions were compliant with the expected constraints, rules, or policies.

Provenance-based *quality assessement* is concerned with the analysis of provenance traces with a view to derive some form of metrics from provenance descriptions. Examples of quality assessement include performance analysis, bottleneck analysis, and user ratings. Quality assessment and compliance checks may be combined, for instance, when regulations require that a system is to be timely, at least 95% of the time.

Provenance-based *cataloging* is the process of building up indexes or lists of artifacts from a provenance trace. For instance, one may list all data sets involved in a computation, or find all agents who bore some responsibility for what happened.

Provenance-based *replaying* can be viewed as a partial re-execution of a provenance trace aimed at deriving novel information about a past computation. Sometimes, re-execution can be total, aiming at *reproducing* computations and results [30].

It is not required to know constraints, rules, policies, analytics metrics, cataloging functions, and replay parameters at application design-time, or even at application runtime. They are checked to be satisfied or computed, at *analysis time*, i.e. the time at which compliance is checked, quality is assessed, catalogs are built, or execution replayed. If analysis takes place after the application execution has completed, it is usually referred to as *offline analysis*, whereas if analyis takes place at execution time, and potentially influences the behavior of the application, then it is referred to as *online analysis*.

Building on Section 2.2, we extract some requirements from the application use cases. Requirements are formulated in terms of a tool that provides some functionality that either is directly required or is similar to functionality required to support the use cases. Requirements cover the four uses of provenance introduced in Chapter 2.

Table 5.1: Requirements categorized by provenance usage

Requirement	Compliance	Quality	Cataloguing	Replay
Policy Check (Requirement 5.5)	✔			
Licensing (Requirement 5.6)	✔			
Timeliness (Requirement 5.7)		✔		
Trust Derivation (Requirement 5.8)		✔		
Credit (Requirement 5.9)			✔	
Confidentiality (Requirement 5.10)				✔
Reproducibility (Requirement 5.11)				✔

5.3.1   PROVENANCE-BASED COMPLIANCE

The first two use cases of Section 2.2.2 are distilled into two requirements pertaining to compliance.

Requirement 5.5   (Policy Check)   The publication policy requires that every article is published after editorial oversight. A policy-check tool determines whether a given article complies with the publication policy.

Online use of the policy-check tool allows the policy to be enforced, ensuring publication of articles only after the publication policy is known to have been complied with. Offline use of the policy-check tools allows auditors to check whether policies are satisfied.

Requirement 5.6   (Licensing)   Given a document, a license tool lists the licenses of all the data sets the document directly or indirectly depends upon and that are published by an external organization.

When publishing open source software, it is a common requirement to list the license of libraries the software depends on. Build tools, such as Maven,[4] support such a requirement[5] by typically relying on an application-specific dependency management mechanism. Requirement 5.6 instead relies on provenance to find associated licenses.

---

[4]http://maven.apache.org
[5]http://mojo.codehaus.org/license-maven-plugin/usage.html

5.3.2   PROVENANCE-BASED QUALITY ASSESSMENT

We extract two requirements from the use cases of Section 2.2.1, related to quality assessment. The first one is concerned with timeliness whereas the second one is about deriving trust.

Very often, a factor to determine the quality of a publication is its timeliness: a publication is timely if it relies on the most recent data available at the time of its publication. This quality assessement is captured by Requirement 5.7.

Requirement 5.7   (Timeliness)   The timeliness-check tool determines that every chart, at the time of its publication in an article, relies on the most recent data set that is available then.

Timeliness is a time-dependent property: a publication may be established to be timely at some point in time, but the release of a new version of the data set this publication relied upon renders it out-of-date. The tool described in Requirement 5.7 can be used online to check timeliness of documents at publication time, or offline, for auditing purposes, in order to assess the performance of a company.

It is part of "provenance folklore" that we put our trust into something because of its provenance. For wine, whisky, marble, respective regions of origin, such as Côte de Nuits, Spey Valley, and Carrara, are very often stamps of quality. It also applies to information and data we find on the Web: we tend to trust news more when it comes from our preferred news outlet. Such a notion of trust, generally subjective, derived from provenance, is captured by the following requirement.

Requirement 5.8   (Trust-Based Filtering)   The trusted-based filtering tool identifies those articles that rely on trusted information sources.

5.3.3   PROVENANCE-BASED CATALOGING

From Section 2.2.3, we derive a requirement pertaining to cataloging. The requirement is intentionally kept simple, so that it can be implemented just by traversing provenance information; in practice, such cataloging operations may require joining provenance with domain specific information.

Requirement 5.9   (Credits)   Given an article, the credits tool builds an acknowledgment list, containing both agents and data sets.

5.3.4   PROVENANCE-BASED REPLAYING

From Section 2.2.4, two further requirements are identified, related to (partial) replaying of an execution using a provenance trace. The first requirement is concerned with a tool that mimics the flow of confidential and non-confidential data in a system.

Requirement 5.10 (Confidentiality)   Given some entities annotated as confidential in a provenance trace, the confidentiality tool determines how confidential data flowed in the application.

The confidentiality tool, used offline, can help detect whether confidential data was inappropriately leaked, and the conditions under which a confidentiality breach may have occurred. Used online, the confidentiality tool can help implement an access control mechanism, granting access to confidential data, according to an organization's prevailing policies.

Requirement 5.11 (Reproducibility)   A reproducibility tool is capable of reproducing a plot, using the same scripts and data sets, or even using variants of scripts or alternate data sets.

Reproducibility is a tenet of the scientific method. While full reproducibility is difficult to achieve (is the hardware still available? have libraries and operating systems changed? have online services and databases been updated?), some partial reproducibility is still very useful, for instance, to reproduce a chart, with a different layout, or in a different format.

5.4   IMPLEMENTATION TECHNIQUES FOR PROVENANCE ANALYSIS

Having discussed use cases and technical requirements for provenance-based tools, we now illustrate how such tools can be implemented using SPARQL queries. We assume the existence of an RDF-based representation of provenance, such as `provapi:d000.ttl`.[6] We rely upon SPARQL 1.1 features such as property paths to implement these queries.

---

[6]In a multi-organizational context such as the data journalism scenario, it is unlikely that such a provenance graph would be readily available for download. In all likelihood, such information is decentralized, since each organization might prefer to independently publish the provenance for the data it produces, such as `provapi:d100.ttl`, `provapi:d200.ttl`, `provapi:d300.ttl`, for NowNews, OtherNews, and PolicyOrg, respectively. Thus, `provapi:d000.ttl` is the kind of provenance that could be scavenged by a consumer of `pol:report1` by taking the union of these individual provenance graphs. The question is: how can those graphs be discovered? In Chapter 4, we have seen techniques by which we can discover that `provapi:d300` denotes the provenance of `pol:report1`. In `provapi:d300`, PolicyOrg has asserted the location of the provenance for `now:employment-article-v1.html`, `other:paper1`, and `other:paper2`. By doing this, PolicyOrg follows the good practice of linking each entity created by an external organization with its respective provenance bundles by means of a new property `bk:topicIn`. After all, before inserting these articles in their report, PolicyOrg did check their provenance. Hence, asserting such `bk:topicIn` properties is not burdensome for publishers, but it is beneficial for consumers, since it allows graphs to be discovered and connected together. As a result, by incrementally loading bundles and following `bk:topicIn`, a consumer is in a position to construct `provapi:d000.ttl`.

5.4.1   FINDING ANCESTORS

When utilizing provenance, a common operation is to trace back all "ancestors" of a resource in a provenance graph. By ancestors, we mean any entity, activity, or agent that occurs in the past of that resource and may somehow have contributed to the way it is. PROV does not specify such a notion of ancestor. Indeed, PROV does not define derivation as a transitive relation, nor does PROV define properties for all potential ancestors: for instance, there is no property to relate an activity to the plan an agent used, or to relate an activity to the starter activity that initiated it, despite the fact that the plan and the latter activity may be seen as ancestors.

Furthermore, we are looking for a permissive notion of ancestor: thus, we include resources that not only have some form of influence on a resource (prov:wasInfluencedBy), but also that are generalizations or alternates of this resource (prov:specializationOf, prov:alternate), or even are contained in the resource (prov:hadMember).

Against this background, we introduce the property bk:ancestor and construct a graph using the construct query form displayed in Figure 5.9. We can assert bk:ancestor between subject ?x and object ?y, if ?x was influenced by ?y, if ?x was a specialization or alternate of ?y, if ?x had ?y as a member, or if ?x is linked to an instance of a qualified class that had a secondary object ?y potentially influencing ?x. This query assumes that reasoning is enabled and that influence, specialization, and alternate properties have been suitably inferred.

5.4.2   DEEP TRAVERSAL

Figure 5.10 displays a SPARQL query that implements Requirement 5.9. It lists all ancestor resources, at any arbitrary depth that are a void:Dataset or a prov:Agent. The query relies on the transitive closure of bk:ancestor, which was defined in Section 5.4.1.

5.4.3   PATTERN DETECTION FOR POLICY COMPLIANCE

A number of compliance requirements can be addressed by searching for graph patterns that exhibit desirable or problematic properties.

Figure 5.11 displays a SPARQL query that implements Requirement 5.5. It returns all resources ?x obtained after publication (these are derivations of type bk:Publication), for which the directly preceding resource ?y was not generated by an activity, with someone in an editorial role, acting according to the publication policy now:editorial-policy. Thus, this query returns problematic resources that do not meet this policy: these are resources with which no activity was documented, for which no editor was associated, or for which the editor did not act according to the expected policy.

The query in Figure 5.11 deals with a simple pattern, directly expressed in SPARQL. While SPARQL is a query language that can deal with sophisticated queries, other approaches may also be considered to handle more complex patterns. Graph theory offers more nuanced approaches, for instance, with algorithms capable of measuring similarity between a pattern and a graph.

```
PREFIX prov: <http://www.w3.org/ns/prov#>
PREFIX bk: <http://www.provbook.org/ns/#>

CONSTRUCT { ?x bk:ancestor ?y. }

WHERE {
  {
    ?x    prov:wasInfluencedBy ?y.
  }
  UNION
  {
    ?x    (prov:specializationOf | prov:alternateOf | prov:hadMember) ?y.
  }
  UNION
  {
    ?x a prov:Activity.
    ?x prov:qualifiedAssociation [ prov:hadPlan ?y ].
  }
  UNION
  {
    ?x a prov:Activity.
    ?x ( prov:qualifiedStart | prov:qualifiedEnd)  [prov:hadActivity ?y].
  }
}
```

Figure 5.9: Asserting ancestors.

5.4.4   TIME COMPARISON

In Figure 5.12, we find a SPARQL query to implement Requirement 5.7 on timeliness. The query finds all images (?y) with a data set (?d) as a primary source, which has been revised into a more recent version (?r). The query retrieves the generation times ?ty, ?td, and ?tr for image ?y, data set ?d, and its revised version ?r, respectively. The image is regarded as not timely if its creation time (?ty) follows that of the data set (?td), and also follows that of the revision (?tr). This indicates that the image was created without making use of a more recent version of the data set.

This query assumes that all clocks used to obtain time information have been synchronized, and that it is safe to perform direct time comparisons.

5.4.5   TRUST-BASED FILTERING

From the outset, we should indicate that it is not our motivation, in this section, to develop new trust models [2]. Data journalists very often regard governmental organizations as the providers of reliable and official data they can base their analysis on. Hence, we adopt a simple approach for trust, according to which a data set is regarded as trusted if it is governmental.

```
PREFIX prov: <http://www.w3.org/ns/prov#>
PREFIX void: <http://vocab.deri.ie/void#>
PREFIX bk: <http://www.provbook.org/ns/#>

SELECT ?x
WHERE {
 <http://www.provbook.org/policyorg/report1>  bk:ancestor* ?x.
 {
   ?x a prov:Entity.
   ?x a void:Dataset.
 }
 UNION
 {
   ?x a prov:Agent.
 }
}
```

Figure 5.10: Credits can be obtained by retrieving all data sets and agents that are ancestors of Poli-cyOrg's report.

```
PREFIX prov: <http://www.w3.org/ns/prov#>
PREFIX bk: <http://www.provbook.org/ns/#>
PREFIX now: <http://www.provbook.org/nownews/>

SELECT ?x ?y

WHERE {
   ?x prov:wasDerivedFrom ?y.
   ?x prov:qualifiedDerivation [ a bk:Publication;
                                 prov:entity ?y ].
   FILTER NOT EXISTS {
      ?y prov:wasGeneratedBy ?a.
      ?a prov:wasAssociatedWith [ a bk:Editor ].
      ?a prov:qualifiedAssociation[ prov:hadPlan  now:editorial-policy]
   }
}
```

Figure 5.11: The publication policy is checked by verifying that any publication had an editorial check according to the publication policy.

Thus, Requirement 5.8 can be addressed by the SPARQL query in Figure 5.13, in which all potential data sets ?y having some influence over an entity contained in a report are returned if they are governmental. The predicate for determining whether a resource is governmental is expressed by the regular expression ^.*://.*gov over the resource's URL.

```
PREFIX prov: <http://www.w3.org/ns/prov#>
PREFIX bk: <http://www.provbook.org/ns/#>
PREFIX void: <http://vocab.deri.ie/void#>
PREFIX schema: <http://schema.org/>

SELECT ?y ?ty ?d ?td ?r ?tr

WHERE {
    <http://www.provbook.org/policyorg/report1> bk:ancestor* ?y.
    ?y a schema:ImageObject.
    ?y prov:qualifiedPrimarySource [ prov:entity ?d].
    ?d a void:Dataset.
    ?r prov:qualifiedRevision       [ prov:entity ?d].
    ?y prov:qualifiedGeneration     [ prov:atTime ?ty].
    ?d prov:qualifiedGeneration     [ prov:atTime ?td].
    ?r prov:qualifiedGeneration     [ prov:atTime ?tr].
}
HAVING  ( ?td < ?ty )  ( ?ty > ?tr )
```

Figure 5.12: Timeliness: The query compares the publication date of a report with the publication date of its primary source and a revised version.

```
PREFIX prov: <http://www.w3.org/ns/prov#>
PREFIX void: <http://vocab.deri.ie/void#>

SELECT ?x ?y

WHERE {
    <http://www.provbook.org/policyorg/report1> prov:hadMember ?x.
    ?x prov:wasInfluencedBy* ?y.
    ?x a prov:Entity.
    ?y a void:Dataset.
    FILTER regex(STR(?y), "^.*://.*gov")
}
```

Figure 5.13: Trust based filtering can be achieved by identifying governmental data sets that have a (direct or indirect) influence on a constituent of a report.

5.4.6 FINDING EXTERNAL ANCESTOR RESOURCES

To address Requirement 5.6, it is useful to identify resources that belong to external organizations. This can be addressed by the SPARQL query in Figure 5.14. This query retrieves all entities that may have directly or indirectly influenced a publication and that have the following characteristics:

either they are attributed to an agent that did not act on behalf of NowNews, or they are not attributed to an agent, but they do not have a NowNews URL.

```
PREFIX prov: <http://www.w3.org/ns/prov#>
PREFIX now: <http://www.provbook.org/nownews/>

SELECT ?x

WHERE {
    <http://www.provbook.org/nownews/employment-article-v1.html> prov:wasInfluencedBy* ?x.
    ?x a prov:Entity.
    {
      ?x prov:wasAttributedTo ?org.
        FILTER NOT EXISTS { ?org prov:actedOnBehalfOf* now:NowNews }
    }
    UNION
    {
        FILTER (!regex(STR(?x), "^http://www.provbook.org/nownews/"))
        FILTER NOT EXISTS { ?x prov:wasAttributedTo ?org }
    }
}
```

Figure 5.14: Licensing Management: This query finds all entities that are attributed to an agent that does not act on behalf of NowNews or that are external and unattributed.

5.4.7   REPLAY TECHNIQUE

Replay techniques consider provenance graphs as a program that can be interpreted, with a view to produce new information. This information can be a reproduction of the original result, or it can be arbitrary metadata, as illustrated in this section.

Figures 5.15 and 5.16 help address Requirement 5.10. They rely on a property bk:hasPrivacy that assigns a level of privacy to its subject when it is an entity, or a level of privacy preservation when it is an activity. For the sake of simplicity, we only consider two possible values for property bk:hasPrivacy:

- :e bk:hasPrivacy bk:Embargo indicates that entity :e is embargoed, meaning that it cannot be shared and is not shared with external organizations before a deadline.

- :a bk:hasPrivacy bk:EmbargoPreserving indicates that activity :a preserves embargoes by not creating outputs before the end of an embargo.

A bk:EmbargoPreserving attribute can be assigned to an activity, for instance, when its underpinning implementation has been certified by some certification authority.

Figure 5.15 displays a query that simply retrieves the confidentiality attribute associated with resources. Figure 5.16 shows the SPARQL query to propagate the confidentiality attribute across derivations. If ?x was derived from ?y and if ?y is confidential, this query helps decide whether ?x is also confidential. The reasoning is as follows. Let us consider the activity ?a that underpinned this derivation and generated ?x. If this activity is not known to be bk:EmbargoPreserving, then one should assume that ?x is an output of ?a that could be generated at any point before the embargo, so it is safe to consider that the embargo rules also apply to ?x, that is, ?x should not be shared with external organizations.

```
PREFIX bk: <http://www.provbook.org/ns/#>

SELECT ?x ?y

WHERE {
    ?x bk:hasPrivacy ?y.
}
```

Figure 5.15: Confidentiality Detection.

```
PREFIX prov: <http://www.w3.org/ns/prov#>
PREFIX bk: <http://www.provbook.org/ns/#>

CONSTRUCT {  ?x bk:hasPrivacy bk:Embargo . }

WHERE {
    ?x a prov:Entity.
    ?x prov:wasDerivedFrom ?y.
    ?y bk:hasPrivacy bk:Embargo.
    FILTER NOT EXISTS {
        ?x prov:qualifiedDerivation ?der.
        ?der prov:entity ?y.
        ?der prov:hadActivity ?a.
        ?a bk:hasPrivacy bk:EmbargoPreserving.
    }
}
```

Figure 5.16: One-Step Confidentiality Propagation: this query sets the confidentiality flag on an activity output, if the activity is non leaking, and its input is confidential.

Given some initial assignment of confidentiality to entities (e.g. an article embargoed till some date) and activities, one can *iteratively* apply the query of Figure 5.16 until no more properties can be inferred. At that point, the provenance graph is annotated with all entities known to be confidential. The iteration is not directly expressed in the query of Figure 5.16. Instead, one has to rely on some program to issue this SPARQL query repeatedly, until no new triples are asserted.

This procedure is known to terminate, since, in any graph, there is a finite number of entities that can be marked as confidential.

Once entities have been marked with the confidentiality flag, further checks can be performed. An auditor can then determine whether access control is properly applied by verifying that no activity used confidential data, unless it was authorized to do so.

5.5   SUMMARY

This chapter began with a presentation of application-independent mechanisms to validate provenance, i.e. determine whether it depicts a consistent history of events amenable to logical reasoning. This was then followed by an introduction to techniques that support further analysis, compliance checks, cataloging, and replaying based on provenance. While such techniques were illustrated by a set of concrete SPARQL queries that contained application-specific structures and parameters, they also exhibited some generic patterns of provenance processing. It is believed that deep traversal, pattern detection, time comparison, trust based filtering, external resource search, and replay techniques are common operations that can be performed on provenance in multiple contexts.

Those techniques are building blocks for more sophisticated functionality, which we refer to as *provenance-based analytics*. Based on provenance, further analysis can lead to predictive models, helping predict future behavior in terms of the past and supporting automated decision making.

CHAPTER 6

Provenance Management

Thus far, we have primarily focused on representing provenance information within systems and, once captured, using that provenance to tackle a variety of use cases. In this chapter, we look at the management of provenance information. In particular, techniques for exposing provenance (Section 6.1) and tools for working with provenance (Section 6.2) are described. We then explain how these techniques, the recipes from Chapter 4, and tools have been deployed on www.provbook.org.

6.1 EXPOSING PROVENANCE

As discussed in Section 4.2, there are two approaches for exposing provenance outlined in the W3C Provenance Access & Query Working Group Note (PROV-AQ) [25]. One is to embed provenance within a data set or document directly. The other is to refer to provenance stored elsewhere. Whether to expose *by value* or *by reference* is use case dependent.

By-value provenance facilitates the provisioning of resources that are self-contained with their provenance, but this has implications on representation size, affecting storage and communications. By-reference provenance allows the sharing of provenance, its updating, and efficient querying, but requires extra communications to dereference it. However, by-value and by-reference provenance are not mutually exclusive. It may be beneficial to embed provenance within a document and also provide a reference to a service that enables querying over that provenance. This is analogous to current practice in the Linked Data community, where information is provided both embedded in content and through SPARQL endpoints. The key approaches for exposing provenance are described in the following recipes described in Section 4.2:

- Embedding Provenance in HTML (Section 4.2.4)

- Embedding Provenance in Other Media (Section 4.2.5)

- Add Provenance to HTTP Headers (Section 4.2.6)

Here, we provide more details on two additional mechanisms for making provenance available: embedding in RDFa and providing provenance services.

6.1.1 EMBEDDING PROVENANCE IN HTML WITH RDFA

Exposing metadata within web pages is increasingly common to support efficient search and indexing. For example, the search engines Google, Bing, Yahoo!, and Yandex started the initiative schema.org to help web developers enrich their pages with additional metadata. A common way of embedding metadata within pages is through the use of RDFa [1]. This section describes how to enrich a web page with PROV using RDFa. We expect readers to have familiarity with HTML and RDFa. The site http://rdfa.info is a good reference for learning about RDFa and tools that use it.

The W3C defined a small subset of RDFa called RDFa-Lite [36] that is designed to be quickly learned and applied. RDFa-Lite predefines a number of popular prefix to namespace bindings for using RDFa, for example, og: for Facebook's Open Graph Protocol and v: for Google's Rich Snippets vocabulary. The PROV namespace is also included in this predefinition using the prefix prov:.

We now walk through marking up the article produced by Bob with the provenance described in Chapter 2. We use RDFa 1.1 for the markup as this allows for expression of datatype information.

Figure 6.1 shows an example of how Bob's employment article now:employment-article-v1.html could look. This is a fairly basic HTML page. The markup is shown in Figure 6.2.

May Employment Report

by Bob created at 2012-03-27T10:00:00-05:00

GovStat has released the May employment report. Included in the report is the mean salary across industries for each state. Below is a chart of the salaries.

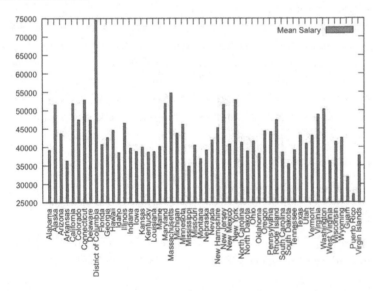

Figure 6.1: Example of Bob's Employment Article now:employment-article-v1.html

```
<!DOCTYPE html>
<html>
    <head>
        <title>May Employment Report</title>
        <link rel="stylesheet" type="text/css" href="nownews.css" />
    </head>
    <body>
        <h2>May Employment Report</h2>
        by Bob 2012-03-27T10:00:00-05:00
        <p>
        GovStat has released the May employment report.
        Included in the report is the mean salary across
        industries for each state.
        Below is a chart of the salaries.
        </p>
        <img  alt="Mean Salary per State" src="plot-v1.jpg"/>
        <hr>
    </body>
</html>
```

Figure 6.2: The HTML of the employment article now:employment-article-v1.html.

To further mark up this page with RDFa, we begin by identifying the entities within the page. This includes the page itself and the image of the plot. In RDFa, it is often necessary to introduce additional structures in order to organize the markup of the page. To refer to the employment report web page itself, a div tag surrounding the main content of the page is introduced. We refer to the page itself using the RDFa resource attribute and define it as being a prov:Entity using the RDFa typeof attribute.

```
....
<div resource="now:employment-article-v1.html"  typeof="prov:Entity">
<h2>May Employment Report</h2>
...
```

A similar approach is used for typing the image. However, instead of adding a surrounding structure, the type is placed directly on the element:

```
<img  alt="Mean Salary per State" src="plot-v1.jpg"  typeof="prov:Entity"/>
```

Notice the introduction of two namespace prefixes now: and prov:. These prefixes need to be defined in the body of the page:

```
<body prefix="prov: http://www.w3.org/ns/prov#
             now: http://www.provbook.org/nownews/">
```

As previously mentioned, we could have omitted the definition of the prov: prefix as it would be recognized by default, but we decided to keep it for pedagogical reasons. Also, in this case, we are making the assumption that the page itself is published under http://www.provbook.org/nownews/, and the image itself is published in the same directory.

Continuing to follow Recipe 4.1.3, the data dependencies can now be described. Here, we simply express that the page itself was derived from the image of the graph (here, derivation is by inclusion of the image in the document). Again, the introduction of an additional structuring element helps to do this:

```
<div resource="now:employment-article-v1.html"  typeof="prov:Entity">
....
 <span rel="prov:wasDerivedFrom">
   <img  alt="Mean Salary per State" src="plot-v1.jpg" typeof="prov:Entity"/>
 </span>
 ...
```

Because of the hierarchy in HTML, an RDFa parser reads this as now:employment-article-v1.html prov:wasDerivedFrom now:plot-v1.jpg. The rel[1] attribute allows us to specify this derivation relationship.

The attribution of the article to Bob can also be expressed again by introducing an additional element, this time specifically referring to an identifier for Bob using the resource attribute. In this additional element, the fact that Bob is a person with responsibility for the article is also stated:

```
<span property="prov:wasAttributedTo"
      resource="now:people/Bob"
      typeof="prov:Person">by Bob</span>
```

The page can further be enriched by encoding information about how the page was produced. To do this, an activity denoting the writing of the article is introduced and connected to the page entity and the data that was used to write the page:

---

[1]Another option is to use the property attribute, but it requires us to include a specific resource attribute to refer to the image. The use of rel means that the parser looks inside the element for the object of the property.

```
<div resource="now:employment-article-v1.html"  typeof="prov:Entity">
   ....
     <span resource="now:is/#writeArticle"
           typeof="prov:Activity">
        <span property="prov:used"
              resource="ftp://ftp.bls.gov/pub/special.requests/oes/oesm11st.zip">
        </span>
         <span property="prov:wasGeneratedBy"
             resource="now:is/#writeArticle">
     </span>
   ....
```

Again, the hierarchy of HTML plays an active role. The statement that
now:employment-article-v1.html   prov:wasGeneratedBy   now:is/#writeArticle
is determined because the prov:wasGeneratedBy statement is within the div containing the
reference to the page itself. This also acts as good example of introducing metadata that is
essentially independent of the displayed HTML. The use of invisible span tags is often helpful to
embed richer provenance information into a web page (e.g. a reference to the data that was used
to create the article).

To complete the example, we introduce the time at which the writing activity ended using the datatype attribute to signify a precise time and linking it to the activity using the
prov:endedAtTime relationship:

```
<span resource="now:is/#writeArticle"
      typeof="prov:Activity">
  <span property="prov:used"
        resource="ftp://ftp.bls.gov/pub/special.requests/oes/oesm11st.zip">
  </span>
  <span property="prov:endedAtTime"
        datatype="xsd:dateTime">2012-03-27T10:00:00-05:00
  </span>
</span>
```

The final marked up page is shown in Figure 6.3. Once the page is marked-up using RDFa
and PROV, the metadata can be easily extracted by a number of tools. Figure 6.4 shows a visualization produced using the RDFa Play tool.[2]

6.1.2   PROVENANCE SERVICES

To manage and query over provenance, it is often beneficial to have a dedicated service or database
specifically for the task. Through the link relation prov:has_query_service, it is possible to

---

[2]RDFa Play: http://rdfa.info/play/

```
<!DOCTYPE html>
<html>
  <head>
    <title>May Employment Report</title>
    <link rel="stylesheet" type="text/css" href="nownews.css" />
    <link rel="http://www.w3.org/ns/prov#has_provenance"
  href="http://www.provbook.org/provapi/documents/d100"/>
  </head>

  <body prefix="prov: http://www.w3.org/ns/prov#
                now: http://www.provbook.org/nownews/">
    <div resource="now:employment-article-v1.html"
           typeof="prov:Entity">
      <h2>May Employment Report</h2>
      <span property="prov:wasAttributedTo"
            resource="now:people/Bob"
            typeof="prov:Person"> by Bob</span> created at
      <span resource="now:is/#writeArticle"
            typeof="prov:Activity">
        <span property="prov:used"
              resource="ftp://ftp.bls.gov/pub/special.requests/oes/oesm11st.zip">
        </span>
        <span property="prov:endedAtTime"
              datatype="xsd:dateTime">2012-03-27T10:00:00-05:00
        </span>
      </span>
      <span property="prov:wasGeneratedBy"
            resource="now:is/#writeArticle">
      </span>

      <p>
        GovStat has released the May employment report.
        Included in the report is the mean salary across
        industries for each state.
        Below is a chart of the salaries.
      </p>
      <span rel="prov:wasDerivedFrom">
        <img alt="Mean Salary per State"
             src="plot-v1.jpg"
             typeof="prov:Entity"/>
      </span>

    </div>
    <hr>
  </body>
</html>
```

Figure 6.3: The provenance of the employment article HTML marked up in RDFa using PROV.

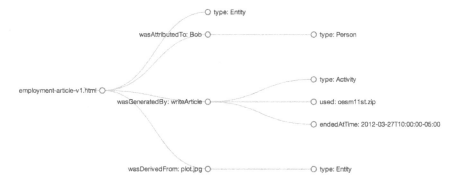

Figure 6.4: RDFa extracted from the employment article and visualized using the RDFa Play tool.

refer to a provenance service that contains the provenance for a given data item. For example, the provenance of the NowNews article could have been provided through a service and specified as follows:

```
<link rel="http://www.w3.org/ns/prov#has_query_service"
      href="http://www.provbook.org/nownews/prov-service"/>
```

To encourage interoperability, PROV-AQ defines a protocol for determining what type of provenance service is at a given URL and the information needed to invoke it. The general procedure for a client retrieving provenance from a provenance service is:

1. retrieve the service description;

2. from the service description, determine the type of provenance service;

3. extract any information necessary to invoke the service;

4. query the service.

For the first step, this is as simple as dereferencing the provenance service URL. A service description should be provided. This can be in any service description language. PROV-AQ provides examples using RDF. Within that description, the client needs to first determine that indeed the service supports provenance. This is done by checking that the service is of type prov:ServiceDescription. Then, the exact type of provenance service can be checked. (When we say "check" this could be as simple as seeing if the target-uri is contained within the description.) PROV-AQ defines two types of provenance services (others could also be defined):

1. A SPARQL Query Service - sd:Service - http://www.w3.org/ns/sparql-service-description#Service

2. Direct HTTP Query Service - prov:DirectQueryService - http://www.w3.org/ns/ prov#DirectQueryService

The reason for these two types of services is that in many cases provenance may be stored in a database or other system that is not RDF based or does not support SPARQL. By defining a simple web service description, such services can move towards interoperability.

Once the type of service is determined, the client needs to extract the information necessary to invoke the service. For a SPARQL Query Service this is essentially the endpoint information as specified by sd:endpoint and a SPARQL query can be issued using the target-uri identified by the prov:has_anchor link relationship. For example, to retrieve the most immediate set of provenance statements for the given target-uri, one could use the following query:

```
select * where {<target-uri> ?p ?o}
```

For the Direct HTTP Query Service, the prov:provenanceUriTemplate needs to be located. This specifies how the target-uri can be passed to the underlying service to retrieve its provenance. For example, for our hypothetical NowNews provenance service, we would provide the following service description:

```
@prefix prov:    <http://www.w3c.org/ns/prov#> .
@prefix now:   http://www.provbook.org/nownews/>

<> a prov:ServiceDescription ;
    prov:describesService now:prov-service.

now:prov-service a prov:DirectQueryService ;
    prov:provenanceUriTemplate "/prov-service?target={+uri}" .
```

This service description specifies that the target-uri should be sent in the target parameter to the web service. The definition of the template should follow the scheme defined by RFC6570.[3] The use of the query parameter ?target is a convention, not a requirement. A benefit to using the Direct HTTP Query Service is that provenance specific parameters can be easily added, such as a parameter to specify the depth of traversal within the provenance graph. Once the invocation information is obtained, the URL for retrieving the provenance information for the given target-uri can be constructed and the service can be invoked. An example URL for retrieving the provenance of the employment article from the described service would be (note the percent encoding of the target-uri):

[3]URI Template: http://tools.ietf.org/html/rfc6570

```
http://www.provbook.org/nownews/prov-service?target=
     http%3A%2F%2Fwww.provbook.org%2Fnownews%2Femployment-article-v1.html.
```

An important point to make is that the mechanisms specified by PROV-AQ are purpose-fully designed to be model and serialization independent. They do not require that provenance be returned in any PROV serialization.[4] For example, a Word or HTML document describing the provenance of a data item could be returned as long as somewhere in the document the specified `target-uri` occurs. This means the exposure mechanisms defined in PROV-AQ can be used in systems that may not wish to fully adopt all of PROV.

6.2   PROVENANCE MANAGEMENT TOOLS

The purpose of this section is to briefly introduce a suite of tools, the *Southampton Provenance Tool Suite*,[5] designed[6] for developers to support the PROV family of specifications. Some of these are libraries in various languages (ProvToolbox is written in Java, ProvPy in Python, and ProvVis in JavaScript), some are command line executables (provconvert), and others are online services (ProvValidator, ProvTranslator, ProvStore).

6.2.1   PROVTOOLBOX

ProvToolbox[7] is a Java toolbox for creating Java representations of the PROV Data Model and manipulating them from the Java programming language. ProvToolbox is capable of marshalling such Java representations to RDF serializations, XML, PROV-N, and PROV-JSON [24] it can also un-marshal such representations back to Java. Furthermore, graphical representations of provenance are possible in the form of SVG (Scalable Vector Graphics) and PDF (Portable Document Format). ProvToolbox is used in other tools, such as provconvert, ProvTranslator, and ProvValidator described in the following sections.

6.2.2   PROVPY

ProvPy[8] provides an implementation of the PROV Data Model in Python. This package provides a memory-based representation of PROV assertions, and offers serialization to and deserialization from PROV-JSON. Persistence is offered by means of the Django Object Relational Mapping. Furthermore, ProvPy offers logging-like facilities to help track provenance in Python programs.

---

[4]Although using PROV serialization is a good idea.
[5]Southampton Provenance Tool Suite: `https://provenance.ecs.soton.ac.uk/`
[6]Trung Dong Huynh is the designer of ProvPy and ProvStore, Danius Michaelides is the designer ProvExtract, and both Danius Michaelides and Alex Fraser are designers of ProvVis.
[7]ProvToolbox: `https://github.com/lucmoreau/ProvToolbox`
[8]ProvPy: `https://pypi.python.org/pypi/prov`

6.2.3 PROVCONVERT AND PROVTRANSLATOR

The binary executable provconvert is a command line executable, based on ProvToolbox. It takes an input file (with possible extensions ttl, rdf, provn, provx, json) and creates an output file (with possible extensions ttl, rdf, provn, provx, json, dot, svg, pdf).

```
provconvert [-infile file] [-help] [-outfile file]
-infile <file>        use given file as input
-help                 print this message
-outfile <file>       use given file as output
```

Alternatively, ProvToolbox is also directly exposed as an online service, ProvTranslator,[9] which allows users to input some PROV representations in various formats, upload a file, or specify a URL so as to convert them into a representation of their choice.

6.2.4 PROVSTORE

Building on ProvPy, ProvStore[10] is a web service that allows users to store, browse, and manage provenance documents. The store can be accessed via a web interface or via a REST API (using an API key or OAuth authentication).

6.2.5 PROVVALIDATOR

ProvValidator[11] is an online service for validating provenance representations according to the PROV-Constraints specification [8]. Figure 6.5, for example, displays the front page of the validation service. The user is invited to submit provenance assertions expressed according to one the PROV representations, a file, or a url for validation. Figure 6.6 displays the validation result for the data journalism example.

The validation result indicates whether all constraints are satisfied; if not, it lists terms that violate constraints. It also displays various warning or errors, such as potentially ambiguous identifiers (in PROV-N or PROV-XML) or syntactically malformed expressions.

A REST API to ProvValidator[12] is also available.

6.2.6 BROWSER PROV EXTRACTOR

ProvExtract[13] is a browser bookmarklet that extracts provenance-related information from the document currently viewed in the web browser. When clicked, a dialog box such as the one illustrated in Figure 6.7 pops up.

The first box therein displays "rel" links appearing in the HTML header, whereas the second box displays any PROV RDFa statements that may be embedded in the document. The user is then

---

[9]ProvTranslator: `https://provenance.ecs.soton.ac.uk/validator/translator.html`
[10]ProvStore: `https://provenance.ecs.soton.ac.uk/store/`
[11]ProvValidator: `https://provenance.ecs.soton.ac.uk/validator/`
[12]Rest API to ProvValidator: `https://provenance.ecs.soton.ac.uk/validator/api.html`
[13]ProvExtract: `https://provenance.ecs.soton.ac.uk/tools/extract/`

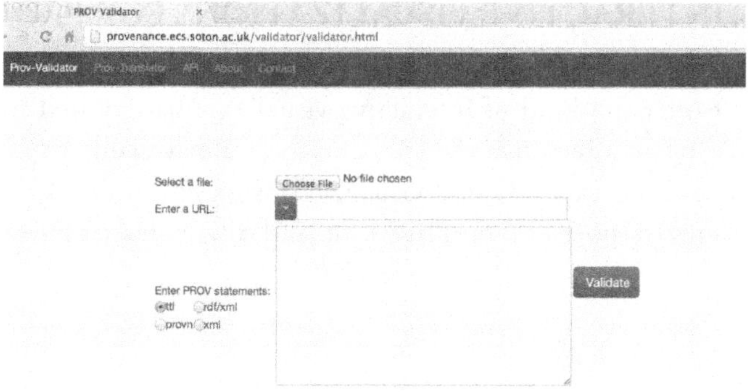

Figure 6.5: Validator Service.

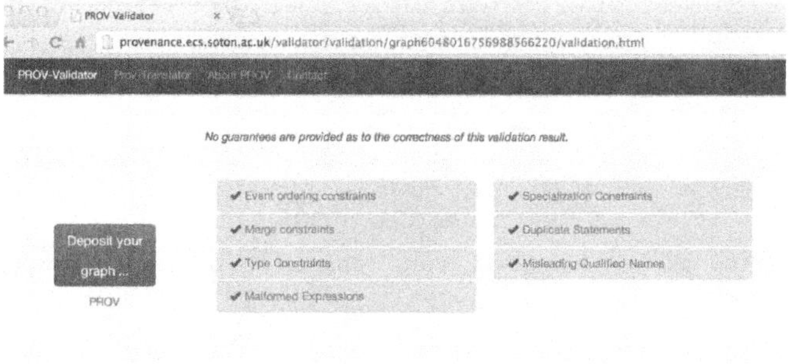

Figure 6.6: Validator Result.

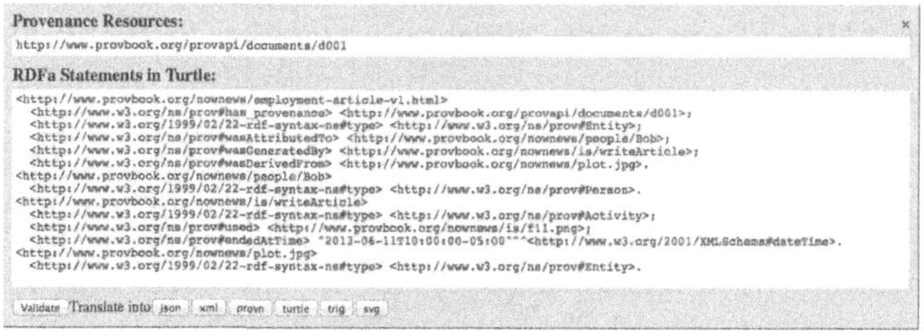

Figure 6.7: Browser PROV Extractor.

offered the options to validate the corresponding provenance, or to translate it to various representations. Validation and translation respectively rely on the ProvValidator and ProvTranslator services.

6.2.7   PROVVIS: INTERACTIVE VISUALIZATIONS FOR PROV

The SVG or PDF renderings of PROV graphs returned by ProvToolbox and ProvTranslator are static, and, therefore, do not offer the kind of interactive visualizations required to explore unknown provenance graphs. Very few interactive visualization tools are available yet to support the PROV standard. ProvVis[14] is a set of experimental visualizations based on D3.js.[15] In this section, we present one of the visualizations, based on Hive Plot [26], illustrated in Figure 6.8.

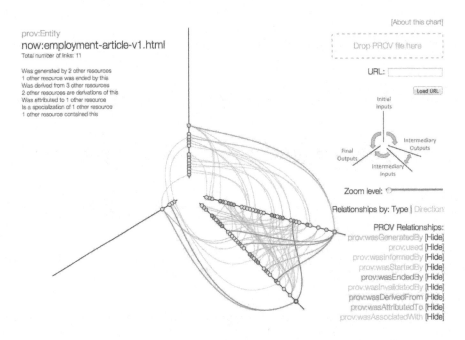

Figure 6.8: Snapshot of Hive Plot interactive visualization tool (see footnote 16) applied to the Data Journalism Example.

In a ProvVis Hive Plot, the PROV elements, i.e. entities, activities, and agents, are grouped according to their involvement in relationships. Elements that occur only as the target of relationships are displayed on the top "spoke"; elements that are only the source of relationships are mapped on the bottom-left "spoke"; finally, elements that are both source and target in some relationships occur in the bottom-right "spokes." So, the top "spoke" contains the resources that do not have ancestors, whereas the bottom-left "spoke" is the place for resources without descendants. On the bottom right, we find intermediary elements, which are replicated on the two right-bottom "spokes" to make the relationships between them clearer.

---

[14]ProvVis: `https://provenance.ecs.soton.ac.uk/vis/`
[15]D3.js: `http://d3js.org/`

Hovering over an element or relationship displays additional information in the top-left of the window. Individual relationship types can be toggled using the selectors at the bottom-right of the chart. The Hive Plot for the data journalism scenario[16] is available for the reader to interact with. In addition to the Hive Plot, two other experimental visualizations are also available: the Wheel plot[17] and the Gantt plot.[18]

6.3 PROVENANCE MANAGEMENT ON WWW.PROVBOOK.ORG

In this section, we explain how we have applied the recipes and techniques presented in this book to manage the provenance at `www.provbook.org`.

6.3.1 DIRECTORIES

Figure 6.11 displays the structure of the `www.provbook.org` website. At the top level, we find four directories `/nownews`, `/gov`, `/othernews`, and `/policyorg` to represent the websites for the four organizations involved in the data journalism scenario, namely NowNews, GovStat, OtherNews, and PolicyOrg. Under each of these directories, we find articles, policies, data sets, and instance stores, where appropriate. For NowNews, we also find a directory for `/people`.

At the toplevel, we also find `/ns` for the namespace used for the book, and `/provapi`, which we respectively discuss in Sections 6.3.3 and 6.3.4.

6.3.2 URI SCHEMES FOR ENTITIES, AGENTS, AND ACTIVITIES

All instances of `prov:Entity`, `prov:Agent`, and `prov:Activity` are identified by a URI. Two distinct URI conventions are adopted.

First, URIs for the persistent characters of the scenario, such as `now:NowNews`, appear in one of the directories described in Section 6.3.1. These URIs denote non-information resources, such as an organization or a person. Content negotiation can be applied to these to obtain TURTLE or HTML representations.

Requesting the HTML representation redirects to the home page for this instance (for instance, `now:NowNews.html`), whereas getting the TURTLE representation results in an RDF description, including the link to the HTML page, with the property `foaf:page`.

```
curl -sH "Accept: text/turtle" -L http://www.provbook.org/nownews/NowNews
@prefix now: <http://www.provbook.org/nownews/> .
@prefix prov: <http://www.w3.org/ns/prov#> .
@prefix schema: <http://schema.org/> .
```

---

[16]Hive Plot: `http://www.provbook.org/provapi/documents/d000/vis/hive`
[17]Wheel Plot: `http://www.provbook.org/provapi/documents/d000/vis/wheel`
[18]Gantt Plot: `http://www.provbook.org/provapi/documents/d000/vis/gantt`

a–little–provenance–goes–a–long–way	28 Jun 2013 13:40	874 bytes	Document
▼ 📁 gov	1 May 2013 06:27	--	Folder
GovStat	1 May 2013 06:27	78 bytes	Document
oes	1 May 2013 06:27	58 bytes	Document
index.html	26 Jun 2013 22:49	356 bytes	HTML...ument
▶ 📁 is	18 Jun 2013 09:38	--	Folder
Luc	18 Jun 2013 09:38	30 bytes	Document
Makefile	12 Apr 2013 14:43	254 bytes	Document
▼ 📁 nownews	Today 12:11	--	Folder
editorial–policy	16 May 2013 07:20	68 bytes	Document
employment–article	16 May 2013 07:20	305 bytes	Document
employment–article–v1.html	28 Jun 2013 07:53	1 KB	HTML...ument
employment–article–v2.html	25 Jun 2013 11:54	213 bytes	HTML...ument
▶ 📁 is	25 Jun 2013 11:54	--	Folder
NowNews	Today 12:08	30 bytes	Document
▼ 📁 people	Today 12:08	--	Folder
Alice	12 Apr 2013 14:43	27 bytes	Document
Bob	12 Apr 2013 14:43	26 bytes	Document
Nick	1 May 2013 06:27	43 bytes	Document
Tom	1 May 2013 06:27	45 bytes	Document
plot.jpg	25 Jun 2013 21:33	154 KB	JPEG image
▶ 📁 ns	28 Jun 2013 07:53	--	Folder
▼ 📁 othernews	16 May 2013 07:20	--	Folder
dataset1	1 May 2013 06:27	38 bytes	Document
dataset2	1 May 2013 06:27	38 bytes	Document
▶ 📁 is	1 May 2013 06:27	--	Folder
OtherNews	1 May 2013 06:27	57 bytes	Document
paper1	16 May 2013 07:20	70 bytes	Document
paper2	16 May 2013 07:20	70 bytes	Document
some–page	16 May 2013 07:20	145 bytes	Document
Paul	18 Jun 2013 09:38	29 bytes	Document
▼ 📁 policyorg	16 May 2013 07:20	--	Folder
▶ 📁 is	1 May 2013 06:27	--	Folder
PolicyOrg	1 May 2013 06:27	135 bytes	Document
report1	16 May 2013 07:20	55 bytes	Document
▶ 📁 provapi	25 Jun 2013 11:54	--	Folder
thebook	28 Jun 2013 09:42	228 bytes	Document

Figure 6.9: Web Site Structure.

```
@prefix foaf: <http://schema.org/> .

now:NowNews a prov:Agent, schema:Corporation;
  foaf:page <http://www.provbook.org/nownews/NowNews.html>.
```

Second, URIs for the ephemeral entities and activities, which never had a presence on the Web during their lifetime, rely on a fragment identifier. For instance, `is:f6` denotes a file on Alice's computer, and `is:unzip2` denotes the activity consisting of decompressing a file. Content negotiation also applies here to get TURTLE or HTML representations.

6.3.3   THE PROV BOOK ONTOLOGY

We have defined an ontology, the PROV book ontology, with namespace `http://www.provbook.org/ns/#` to capture various concepts related to the data journalism scenario. The ontology specifies a few classes and properties structured along a class hierarchy displayed in Figure 6.10. The PROV book ontology extends the PROV ontology, by subclassing `prov:Activity`, `prov:Derivation`, `prov:Plan`, and `prov:Role`.

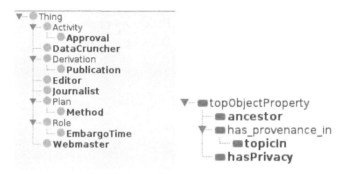

Figure 6.10: PROV Book Ontology Structure. Classes and properties defined in this book appear in bold.

Content negotiation is also supported on resource `http://www.provbook.org/ns/`, with TURTLE and HTML among the formats supported.

6.3.4   DATA JOURNALISM PROVENANCE

All the provenance is exposed at `/provapi/documents`. Figure 6.11 summarizes the various provenance documents available at this location. The first set (`provapi:d000–provapi:d003`) consists of examples presented in Chapter 4. The second set (`provapi:d100, provapi:d200, provapi:d300`) contains provenance produced by NowNews, OtherNews, and PolicyOrg, respectively. Finally, `provapi:bk` is concerned with provenance of material of this book.

Each of these documents is a non-information resource, for which, content negotiation is applicable. When a bundle is exposed, given that the TURTLE and RDF/XML representations do not contain the named graph representing the bundle, other representations are preferable. Hence, the quality is set to 0.5 (qs=0.5).

6.3.5   EXPOSING PROVENANCE

For each resource on `www.provbook.org`, one can discover the provenance for that resource, using the following techniques:

- All HTML, JPG, SVG documents contain provenance metadata, according to recipes in Section 4.2.4 and 4.2.5.

URI	Description
provapi:d000 provapi:d001 provapi:d002 provapi:d003	Pedagogical Material
provapi:d100 provapi:d200 provapi:d300	NowNews provenance OtherNews provenance PolicyOrg provenance
provbook:provenance alias of provapi:bk	Book provenance

Figure 6.11: Provenance at /provapi/documents.

```
curl -s -D - http://www.provbook.org/provapi/documents/d300 -o  /dev/null -H "Accept: foo"

HTTP/1.1 406 Not Acceptable
Date: Mon, 01 Jul 2013 14:21:53 GMT
Server: Apache/2.2.3 (Red Hat)
Alternates: {"d300.provn" 1 {type text/provenance-notation} {length 1763}},
            {"d300.provx" 1 {type application/provenance+xml} {length 3118}},
            {"d300.ttl" 0.5 {type text/turtle} {length 1748}},
            {"d300.rdf" 0.5 {type application/rdf+xml} {length 3323}},
            {"d300.json" 1 {type application/json} {length 3443}},
            {"d300.svg" 1 {type image/svg+xml} {length 10252}},
            {"d300.jpg" 1 {type image/jpg} {length 33070}},
            {"d300.trig" 1 {type application/trig} {length 1793}}
Vary: negotiate,accept
TCN: list
Content-Length: 846
Connection: close
Content-Type: text/html; charset=iso-8859-1
```

Figure 6.12: Choice of Format and Quality Available For Content Negotiation.

- Some HTML documents also contain by-value provenance expressed in RDFa.

Furthermore, for all resources served by www.provbook.org, the HTTP Link Header is set to point to provbook:provenance, a convenience URL for provapi:bk. For instance, provbook:thebook is associated with the link header displayed in Figure 6.13.

Additionally, resources related to the data journalism application have provenance from two different perspectives. All these resources were created by the authors of this book, but they also have some provenance in the context of the application.

```
curl -s -D - http://www.provbook.org/thebook -o  /dev/null

HTTP/1.1 200 OK
Date: Mon, 01 Jul 2013 22:54:39 GMT
Server: Apache/2.2.3 (Red Hat)
Last-Modified: Fri, 28 Jun 2013 10:27:36 GMT
ETag: "1298d58-e4-4e03453d17600"
Accept-Ranges: bytes
Content-Length: 228
Link: <http://www.provbook.org/provenance>;
        rel="http://www.w3.org/ns/prov#has_provenance";
        anchor="http://www.provbook.org/thebook"
Connection: close
Content-Type: text/html; charset=UTF-8
```

Figure 6.13: Single Link Header.

Hence, `provapi:d300` is a self-referential bundle, and also has provenance in `provbook:provenance`. This is reflected in the multiple link headers displayed in Figure 6.14.

```
curl -s -D - http://www.provbook.org/provapi/documents/d300 -o /dev/null -H "Accept: text/turtle"

HTTP/1.1 200 OK
Date: Mon, 01 Jul 2013 14:24:39 GMT
Server: Apache/2.2.3 (Red Hat)
Content-Location: d300.ttl
Vary: negotiate,accept
TCN: choice
Last-Modified: Mon, 01 Jul 2013 12:27:13 GMT
ETag: "3738030-6d4-4e072591fea40;18b849b-18f-4e073f24f8080"
Accept-Ranges: bytes
Content-Length: 1748
Link: <http://www.provbook.org/provapi/documents/d300>;
        rel="http://www.w3.org/ns/prov#has_provenance";
        anchor="http://www.provbook.org/provapi/documents/d300.ttl",
      <http://www.provbook.org/provenance>;
        rel="http://www.w3.org/ns/prov#has_provenance";
        anchor="http://www.provbook.org/provapi/documents/d300.ttl"
Connection: close
Content-Type: text/turtle
```

Figure 6.14: Multiple Link Headers.

This example shows that multiple provenance "accounts" for a given resource can co-exist. They provide different perspectives, which users may decide to use according to their need.

6.4   SUMMARY

In this chapter, we have presented two mechanisms for exposing provenance information by either embedding that information within web pages or by providing services that enable its retrieval. We also discussed the Southampton Provenance Tool Suite—a series of tools for storing, manipulating, retrieving, and visualizing provenance. Beyond what was presented here, there are a variety of applications and frameworks that generate or help manage provenance. At the time of the publication of the PROV recommendations, there were 41 reported implementations [23] and we hear of more every day.

CHAPTER 7

Conclusion

We have now completed our introduction to PROV. We have surveyed the PROV-O ontology and presented a set of recipes that allow developers to model their application, and organize provenance in a way that it can be discovered and processed in an interoperable manner. We have outlined the principles of validation, to encourage developers to produce provenance that is "correct." We have examined various ways of utilizing provenance, allowing us to address the use cases we have introduced, in relationship to the data journalism scenario. The online resources at www.provbook.org provide a concrete example of how a website can be completely provenance-enabled. In lieu of a summary, we present a checklist of steps to ensure provenance is structured and exposed adequately. We then look at open issues pertaining to provenance.

7.1 TOWARD PROVENANCE SELF CERTIFICATION: A CHECKLIST

Having read the book and applied the techniques to a given application, a developer may wonder what good provenance looks like. Figure 7.1 displays a checklist that provides developers with guidance to assess the quality of their provenance. Over time, we may expect this checklist to become the basis of a self-certification process, akin to open data self-certification.[1]

Figure 7.1 displays the *good provenance checklist*. The first item principally focuses on the contents of the provenance, whereas the remaining items are more concerned with the way it is exposed and made accessible.

As far as point (1f) in Figure 7.1 is concerned, the PROV family of specifications offers a set of pre-defined types:

- When appropriate, use PROV subtypes for entities and agents, such as prov:Plan, prov:Bundle, prov:Person, prov:SoftwareAgent, and prov:Organization.

- When suitable, use PROV-DC subtypes for entities and activities, such as dct:Policy, prov:Publish, prov:Modify, and prov:Replace [11].

- When appropriate, use PROV subtypes for derivations, such as prov:wasRevisionOf, prov:hadPrimarySource, and prov:wasQuotedFrom.

- When suitable, use PROV-DC subtypes for derivations, such as dct:isFormatOf, dct:source, and dct:references.

[1]Open data certificate: https://certificates.theodi.org/

- When suitable, use PROV-DC-defined subtypes for attribution, such as dct:creator, dct:contributor, dct:publisher, and dct:rightsHolder.

---

1. Encode provenance, by successively expressing data flow, responsibility, process, and alternate views.

   (a) Identify entities, activities, and agents in all their facets, versions, or instances.

   (b) Express data flow by relating entities with derivation properties.

   (c) Assign responsibilities for entities and other agents.

   (d) Refine derivations and responsibilities by describing activities and by annotating events (prov:InstatenousEvent) with time information.

   (e) Relate entities with specialization and alternate properties.

   (f) Refine provenance by subtyping entities, activities, agents, and properties.

2. Bundle up provenance and express provenance of provenance.

3. Make provenance web-accessible, after setting a suitable access control policy; enable content negotiation to allow clients to select the serialization format of provenance.

4. Embed provenance metadata in documents, in data, and in general application resources. If provenance metadata cannot be embedded, embed this information in HTTP headers.

5. Validate provenance and check that provenance can be retrieved for any resource.

Figure 7.1: Good Provenance Checklist.

7.2 APPLYING PROVENANCE IN THE WILD

Chapter 1 introduced several application areas that were particularly conducive to the use of provenance: food supply chain, open data and data journalism, traceability in the Social Web, reproducibility of science, accountability, transparency, and compliance in business applications. All have requirements similar to those discussed in Chapter 5.

From a technical perspective, notwithstanding the open issues discussed in Section 7.3, PROV provides a lightweight vocabulary that can be incrementally deployed to tackle real application requirements. A challenge that is likely to emerge in most application domains is the need for common vocabulary specializing PROV: further standardization activities may help develop community consensus for various domains.

For most novel technologies, human factors may cause resistance to change. It may well be the case for provenance too. Some individuals or organizations may fear that provenance could expose bad practice or that provenance may reveal a business' "secret sauce." It is therefore crucial to determine what the *incentives* of recording, storing, and sharing provenance are for a given application. An organization would have to calculate the return on investment of provenance. As some businesses in the food sector demonstrate, provenance can be a competitive advantage, forming an integral part of the branding of a product, alleviating the fear of transparency that provenance might carry for some.

7.3 OPEN ISSUES

Standardization for provenance is not the end of the road. Far from it, it is an enabler for adoption, integration with multiple systems across the Web, and its potentially widespread use. The W3C Provenance Working Group, despite releasing some 12 specifications, had a limited charter, and simply did not tackle all the issues related to provenance. We discuss some of them in this section.

7.3.1 PROVENANCE ENABLING SYSTEMS

Tools for Provenance

In order to bootstrap a provenance transparent Web, a crucial element is to be able to develop new applications that generate provenance. Hence, a line of research is libraries and language bindings to allow programming language runtimes to generate provenance automatically. Techniques such as aspect-oriented programming and reflection may come useful. Chapter 6 presented emerging efforts, in the form of libraries and services, in that space.

There already exist applications that record some form of provenance, though they do not expose it as PROV. For instance, word processors, wikis, and version control systems can keep a record of all the versions of documents they generate. Conversion tools converting their proprietary internal representations to PROV would allow this provenance to be exposed in an interoperable manner.

Ultimately, provenance needs to be stored and processed. Systems for storing, querying, and processing provenance, possibly relying on triple stores and graph databases, need to be designed secure and scalable. APIs needs to be specified and standardized to interact with such systems.

Legacy Systems: Provenance Reconstruction

We need to be able to integrate with legacy systems: in this context, these are systems that do not record any form of provenance and that we cannot redesign and rewrite. Hence, the techniques discussed above (enrich programming languages with provenance capabilities and convert existing provenance) are not applicable to such legacy systems. Instead, provenance needs to be *reconstructed* from existing data generated by, and from knowledge about, these systems. Provenance reconstruction may need to rely on data mining techniques.

Provenance Methodology

The provenance recipes presented in Chapter 4 are an initial attempt at defining a methodology for deploying PROV. Several methodological facets are worth studying. *(i)* A methodology is required to identify which entities, activities, and agents in an application need to be represented in provenance traces so as to support provenance requirements. PRIME [28], a methodology defined in the context of OPM [31], is a good starting point that could serve as the basis for a PROV-specific methodology. *(ii)* Provenance design patterns would enrich this methodology and increase interoperability. This book's recipes provides the first set of such patterns; their usage should be documented in the methodology. *(iii)* Provenance, like security, should not be a last-minute add-on to a system. Provenance should be integrated in the design phase, and support for it should exist in design tools and IDEs.

7.3.2 FUNDAMENTALS OF PROVENANCE

It is an open question as to whether a "grand unified theory" of provenance can be developed to join up the various theoretical foundations emerging in various communities, for PROV [6], OPM [30], databases [7], and linked data [9].

The W3C Provenance Working Group has defined concepts that were more speculative, and their interpretation in the context of a theoretical framework would help improve their understanding: for instance, the lack of theoretical foundations and unambiguous meaning for prov:Bundle, prov:mentionOf [32], and prov:Dictionary may hamper their understanding, and therefore, their usage; furthermore, potential useful opportunities to reason about these constructs may be missed.

Alternative data models of provenance, such as provenance models with probabilistic weights, and provenance interpretation based on causality would provide a less discrete and less deterministic view of provenance, which would suit contexts in which provenance information is uncertain or partially known. We note that PROV was designed to support such models through its extension mechanisms.

7.3.3 PROVENANCE ANALYTICS

As discussed in Chapter 1, the provenance of data enables trust judgment to be derived about such data. It is desirable for trust judgments to be automatically computable since humans would very quickly be overwhelmed by the quantity of information to processs. Hence, an interesting direction of research could focus on machine learning techniques capable of deriving trust, and, in general, capable of classifying entities or agents, based on provenance. Additionally, providing the equivalent of a "fair trade" stamp of approval based on trust judgments is an interesting path forward [14].

Provenance describes past events that led to some activity or entity. If this description faithfully describes a system, then it can be used as a way of auditing the system, and checking whether the system behaved as expected. Norms, policies, or regulations can specify the expected behavior of a system. Compliance aims to determine whether provenance is compatible with such norms, policies, or regulations. Furthemore, provenance could be used as a predictive model, helping predict next steps of a system.

7.3.4   SECURING PROVENANCE

A data product and its provenance may have different sensitivity [5] and, therefore, need to be handled with appropriate access control [29]. In an employee's performance review, the data product is the review itself, which is available to the employee; its provenance encompasses the authors of the review, who may have to remain anonymous. Hence, the employee can see the data but not its provenance. Symmetrically, a job applicant typically provides the names of the references (and sometimes the actual reference in a sealed envelope); the reference is to remain invisible to the applicant, while its provenance is known to the applicant.

An assumption generally made is that provenance is faithful and accurately reflects past executions. Such properties can be enforced by cryptographic means, but no standard way of signing provenance traces has been specified yet. Assuming that provenance enjoys a property of integrity, it is also important to ensure that there exists an unforgeable link between provenance and the data it is about.

7.4   FINAL WORDS

PROV is a rich vocabulary that was designed to tackle a whole variety of use cases. Importantly, PROV's design is such that its adoption hurdle is minimal. Echoing Jim Hendler,[2] we argue that:

A little provenance goes a long way.

Simply identifying a resource, exposing its authors with attribution, and expressing what it is derived from it goes a long way toward a provenance-enabled Web, which comes with all the benefits of transparency, credits, and accountability. This motto can be summed up by the following statements.

---

[2]A little semantics: http://www.cs.rpi.edu/~hendler/LittleSemanticsWeb.html

```
provbook:a-little-provenance-goes-a-long-way a prov:Entity;
   prov:value "A little provenance goes a long way";
   prov:wasAttributedTo provbook:Paul, provbook:Luc;
   prov:wasDerivedFrom <http://www.cs.rpi.edu/~hendler/LittleSemanticsWeb.html>.
```

Finally, for discussions, updates, and more information about material presented in this book, the reader is invited to visit the provbook blog at http://blog.provbook.org.

Bibliography

[1] Ben Adida, Ivan Herman, Manu Sporny, and Mark Birbeck. RDFa 1.1 Primer. Rich Structured Data Markup for Web Documents. Technical report, World Wide Web Consortium, 2012. 82

[2] Donovan Artz and Yolanda Gil. A survey of trust in computer science and the semantic web. *Journal of Web Semantics*, 5(2):58–71, 2007. DOI: 10.1016/j.websem.2007.03.002. 74

[3] Tim Berners-Lee. Semantic Web - XML2000. Invited talk at XML 2000. 4

[4] Tim Berners-Lee and Mark Fischetti. *Weaving the Web: The Original Design and Ultimate Destiny of the World Wide Web by Its Inventor.* Harper San Francisco, 1st edition, 1999. 2

[5] Uri Braun, Avraham Shinnar, and Margo Seltzer. Securing provenance. In *HOTSEC'08: Proceedings of the 3rd conference on Hot topics in security*, pages 1–5, Berkeley, CA, USA, 2008. USENIX Association. 103

[6] James Cheney. Semantics of the PROV Data Model. W3C Working Group Note NOTE-prov-sem-20130430, World Wide Web Consortium, April 2013. 102

[7] James Cheney, Laura Chiticariu, and Wang-Chiew Tan. Provenance in databases: Why, how, and where. *Foundations and Trends in Databases*, 1(4):379–474, 2009. DOI: 10.1561/1900000006. 102

[8] James Cheney, Paolo Missier, Luc Moreau (eds.), and Tom De Nies. Constraints of the PROV Data Model. W3C Recommendation REC-prov-constraints-20130430, World Wide Web Consortium, October 2013. 5, 6, 61, 62, 63, 66, 90

[9] Mariangiola Dezani, Ross Horne, and Vladimiro Sassone. Tracing where and who provenance in linked data: a calculus. *Theoretical Computer Science*, to appear, May 2012. DOI: 10.1016/j.tcs.2012.06.020. 102

[10] Leigh Dodds and Ian Davis. *Linked Data Patterns: A pattern catalogue for modelling, publishing, and consuming Linked Data.* 2012. 23

[11] Daniel Garijo, Kai Eckert (eds.), Simon Miles, Craig M. Trim, and Michael Panzer. Dublin Core to PROV Mapping. W3C Working Group Note NOTE-prov-dc-20130430, World Wide Web Consortium, April 2013. 43, 99

[12] Yolanda Gil, Ewa Deelman, Mark Ellisman, Thomas Fahringer, Geoffrey Fox, Dennis Gannon, Carole Goble, Miron Livny, Luc Moreau, and Jim Myers. Examining the challenges of scientific workflows. *IEEE Computer*, 40(12):26–34, December 2007. DOI: 10.1109/MC.2007.421. 3

[13] Jonathan Gray, Liliana Bounegru, and Lucy Chambers, editors. *Data Journalism Handbook 1.0 BETA*. O'Reilly Media, July 2012. 2, 9

[14] Paul Groth. Transparency and reliability in the data supply chain. *IEEE Internet Computing*, 17(2):69–71, 2013. DOI: 10.1109/MIC.2013.41. 103

[15] Paul Groth, Yolanda Gil, James Cheney, and Simon Miles. Requirements for provenance on the web. *International Journal of Digital Curation*, 7(1), 2012. DOI: 10.2218/ijdc.v7i1.213. 9

[16] Paul Groth, Simon Miles, and Luc Moreau. Preserv: Provenance recording for services. In *UK e-Science All Hands Meeting 2005*. EPSRC, 2005. Event Dates: September 2005. 56

[17] Paul Groth, Simon Miles, and Luc Moreau. A Model of Process Documentation to Determine Provenance in Mash-ups. *Transactions on Internet Technology (TOIT)*, 9(1):1–31, 2009. DOI: 10.1145/1462159.1462162. 48

[18] Paul Groth and Luc Moreau. Representing distributed systems using the open provenance model. *Future Generation Computer Systems*, 27(6):757 – 765, 2011. DOI: 10.1016/j.future.2010.10.001. 45

[19] Paul Groth and Luc Moreau (eds.). PROV-Overview. An Overview of the PROV Family of Documents. W3C Working Group Note NOTE-prov-overview-20130430, World Wide Web Consortium, April 2013. 5

[20] James Hendler. COMMUNICATION: Enhanced: Science and the Semantic Web. *Science*, 299(5606):520–521, 2003. DOI: 10.1126/science.1078874. 3

[21] A J G Hey and A E Trefethen. The data deluge: An e-science perspective. In *Grid Computing - Making the Global Infrastructure a Reality*. Wiley and Sons, 2003. DOI: 10.1002/0470867167.ch36. 3

[22] Hook Hua, Curt Tilmes, Stephan Zednik (eds.), and Luc Moreau. PROV-XML: The PROV XML Schema. W3C Working Group Note NOTE-prov-xml-20130430, World Wide Web Consortium, April 2013. 6

[23] Trung Dong Huynh, Paul Groth, and Stephan Zednik (eds.). PROV Implementation Report. W3C Working Group Note NOTE-prov-implementations-20130430, World Wide Web Consortium, April 2013. 98

[24] Trung Dong Huynh, Micahel Jewell, Amir Sezavar Keshavarz, Danius Michaelides, Huan-jia Yang, and Luc Moreau. The PROV-JSON Serialization. A JSON Representation for the PROV Data Model. Technical report, University of Southampton, 2013. 89

[25] Graham Klyne, Paul Groth (eds.), Luc Moreau, Olaf Hartig, Yogesh Simmhan, James My-ers, Timothy Lebo, Khalid Belhajjame, and Simon Miles. PROV-AQ: Provenance Access and Query. W3C Working Group Note NOTE-prov-aq-20130430, World Wide Web Consortium, April 2013. 6, 51, 81

[26] Martin Krzywinski, Inanc Birol, Steven JM Jones, and Marco A Marra. Hive plots— rational approach to visualizing networks. *Briefings in Bioinformatics*, 2011. DOI: 10.1093/bib/bbr069. 92

[27] Timothy Lebo, Satya Sahoo, Deborah McGuinness (eds.), Khalid Behajjame, James Ch-eney, David Corsar, Daniel Garijo, Stian Soiland-Reyes, Stephan Zednik, and Jun Zhao. PROV-O: The PROV Ontology. W3C Recommendation REC-prov-o-20130430, World Wide Web Consortium, October 2013. 5, 6, 21

[28] Simon Miles, Paul Groth, Steve Munroe, and Luc Moreau. PrIMe: A methodology for developing provenance-aware applications. *ACM Transactions on Software Engineering and Methodology*, June 2009. DOI: 10.1145/2000791.2000792. 39, 102

[29] Luc Moreau. The foundations for provenance on the web. *Foundations and Trends in Web Science*, 2(2–3):99–241, November 2010. DOI: 10.1561/1800000010. 103

[30] Luc Moreau. Provenance-based reproducibility in the semantic web. *Web Semantics: Science, Services and Agents on the World Wide Web*, 9:202–221, February 2011. DOI: 10.1016/j.websem.2011.03.001. 69, 102

[31] Luc Moreau, Ben Clifford, Juliana Freire, Joe Futrelle, Yolanda Gil, Paul Groth, Natalia Kwasnikowska, Simon Miles, Paolo Missier, Jim Myers, Beth Plale, Yogesh Simmhan, Eric Stephan, and Jan Van den Bussche. The open provenance model core specifi-cation (v1.1). *Future Generation Computer Systems*, 27(6):743–756, June 2011. DOI: 10.1016/j.future.2010.07.005. 102

[32] Luc Moreau and Timothy Lebo. Linking across provenance bundles. W3C Working Group Note NOTE-prov-sem-20130430, World Wide Web Consortium, April 2013. 102

[33] Luc Moreau, Paolo Missier (eds.), Khalid Belhajjame, Reza B'Far, James Cheney, Sam Coppens, Stephen Cresswell, Yolanda Gil, Paul Groth, Graham Klyne, Timothy Lebo, Jim McCusker, Simon Miles, James Myers, Satya Sahoo, and Curt Tilmes. PROV-DM: The PROV Data Model. W3C Recommendation REC-prov-dm-20130430, World Wide Web Consortium, October 2013. 6, 21

[34] Luc Moreau, Paolo Missier (eds.), James Cheney, and Stian Soiland-Reyes. PROV-N: The Provenance Notation. W3C Recommendation REC-prov-n-20130430, World Wide Web Consortium, October 2013. 6

[35] Laney Salisbury and Aly Sujo. *Provenance — How a con man and a forger rewrote the history of modern art*. The Penguin Press, 2009. 1

[36] Manu Sporny. RDFa Lite. W3C Recommendation, World Wide Web Consortium, 2012. 82

[37] Brian Tierney, William Johnston, Brian Crowley, Gary Hoo, Chris Brooks, and Dan Gunter. The netlogger methodology for high performance distributed systems performance analysis. In *Proceedings of the 7th IEEE International Symposium on High Performance Distributed Computing*, HPDC '98, pages 260–, Washington, DC, USA, 1998. IEEE Computer Society. DOI: 10.1109/HPDC.1998.709980. 56

[38] Daniel J. Weitzner, Harold Abelson, Tim Berners-Lee, Joan Feigenbaum, James Hendler, and Geral Jay Sussman. Information accountability. *Communications of the ACM*, 51(6):81–87, June 2008. DOI: 10.1145/1349026.1349043. 3

Authors' Biographies

PROFESSOR LUC MOREAU

Professor Luc Moreau is Deputy Head (Research and Enterprise) of Electronics and Computer Science (ECS), and a member of the Web and Internet Science (WAIS) group, at the University of Southampon. He has a long-standing interest in large-scale, open, distributed systems. More recently, he initiated the field of provenance in distributed heterogeneous systems. After launching the International Provenance and Annotation Workshop (IPAW) series in 2006, he instigated and led the highly successful Provenance Challenge, an activity that took place three times, to investigate the interoperability of provenance systems. This has involved over 20 teams across the world, from academia and industry, including a mix of disciplines from end users to technologists. It resulted in the Open Provenance Model (OPM) specification, a model with a growing community adoption, and a key driver for a follow-on standardization activity. Then, Luc Moreau co-chaired the W3C Provenance Working Group, and co-edited three recommendations related to the W3C PROV provenance model. He is currently co-investigator of three large-scale multi-site projects Orchid (`http://www.orchid.ac.uk`), Sociam (`http://www.sociam.org`), and SmartSociety (`http://www.smart-society-project.eu`).

DR. PAUL GROTH

Paul Groth is an assistant professor in the Web & Media Group at the VU University of Amsterdam and a member of its Network Institute. He holds a Ph.D. in Computer Science from the University of Southampton (2007) and has done research at the University of Southern California. His research focuses on dealing with large amounts of diverse contextualized knowledge with a particular focus on the web and e-Science applications. This includes research in data provenance, web science, knowledge integration and knowledge sharing. He has over over 60 publications in these areas. Paul is an active member of the Semantic Web and provenance communities serving on numerous program and organization committees. Paul co-chaired the W3C Provenance Working Group. Currently, he is lead architect of Open PHACTS (`http://www.openphacts.org`) - a project to develop a provenance-enabled platform for pharmacological data integration. He blogs at `http://thinklinks.wordpress.com`.

Index

Printed in the United States
by Baker & Taylor Publisher Services